Law and Policy in Petroleum Development

Law and Policy in Petroleum Development

Changing Relations between Transnationals and Governments

A comparative study sponsored by the Commonwealth Secretariat

Kamal Hossain

Frances Pinter (Publishers) Ltd London
Nichols Publishing Company New York

318020

First published in Great Britain by
Frances Pinter (Publishers) Ltd.
5 Dryden Street, London WC2E 9NW

ISBN 0 903804 43 3

Published in the U.S.A. in 1979 by
Nichols Publishing Company
Post Office Box 96
New York, N.Y. 10024

Library of Congress Cataloging in Publication Data

Hossain, Kamal.
    Law and policy in petroleum development.

        Bibliography: p.
        1. Underdeveloped areas – Petroleum law and legislation.
2. Petroleum law and legislation.   3. International
business enterprises – Law and legislation.   I. Title.
K3915.4.H67   343.07'7   79–1905
 ISBN 0–89397–056–5

Printed in Great Britain by A. Wheaton & Co. Ltd., Exeter

Contents

Chapter III
**Government and Petroleum Development:
Objectives, Options and Strategies** 58

vi

Preface

The present comparative study of law and policy relating to petroleum development was undertaken as a research project by the author under the terms of the Research Fellowship to which he was elected by Nuffield College, Oxford, with effect from October 1975. The author's interest in this area grew out of his experience as a member of a developing country government (Bangladesh), when he had been involved in negotiations with oil companies, relating to off-shore exploration and other matters. It had been evident at that time that government negotiators were handicapped by lack of information of recent developments in the field of law and policy relating to petroleum development.

Under the impact of major developments, significant changes had been taking place in the relations between governments and transnational oil companies. The doctrine of permanent sovereignty over natural resources propounded in the UN General Assembly in the sixties gained heightened relevance in the context of the post-1973 global energy situation. The doctrine was recognised as a basic principle of the new international economic order by the Declaration adopted at the Sixth Special Session of the UN General Assembly (May 1974) and in the Charter of Economic Rights and Duties of States (December 1974). Implicit in such recognition was the determination of governments to secure effective control over their natural resources, leading to fundamental policy reappraisals not only in the established oil-producing (OPEC) countries, but also in the new producing countries, such as Britain and Norway.

These developments had special relevance for countries, in particular developing countries, which were poised to embark on programmes of petroleum exploration. A basic aim of the present study is to provide some guidance and information about relevant developments in this field to developing country legislators, policy-makers, administrators, and negotiators.

The comparative study has concentrated on a number of selected countries — Australia, Bangladesh, Canada, India, Indonesia, Malaysia, New Zealand, Norway, Trinidad and Tobago, and Venezuela. The choice of countries was guided in part by the desire to make a representative selection of different approaches, and in part by the availability of material. In addition to examination of published materials, including texts of agreements and legislation, questionnaires were sent to a number of the governments, and interviews were held with government and company representatives in Britain, Canada, Indonesia, Norway, the United States of America and Venezuela.

The author gratefully acknowledges valuable assistance received from institutions and persons, too numerous to be named individually. Some of them, however, must be singled out for specific mention because of the continuity assistance received from them throughout the project. A profound expression of gratitude must therefore be extended to: Nuffield College, Oxford, for electing the author to a Research Fellowship and for providing the most congenial arrangements for work, to its Warden during the relevant period, Sir Norman Chester, who from beginning to end offered assistance and encouragement, as well as invaluable comments on the draft, which despite his many preoccupations he found the time to read, and to the author's colleagues, Mr Maurice Scott, Mr Aubrey Silberston and Mr Laurence Whitehead, Fellows of Nuffield College, for their most useful suggestions and comments; the Commonwealth Secretariat and its Secretary-General, Mr. S. S. Ramphal and Deputy Secretary-General during the relevant period, Mr M. Azim Husain, for encouragement and support extended to the project, and to the Commonwealth Fund for Technical Co-operation for the research grant made to the College, which made the project possible, and its officials, Mr Brian Tyler, Mr Roland Brown, Mr Arthur Hazelwood, Mr Lindsay Smallbone, Mr Peter Freeman, and Mr Maqbul Rahim, for their assistance at different stages, including valuable comments on the draft; the Chr. Michelsen Institute, Norway, and the Director of its Development Research Programme, Dr Just Faaland and his associate, Mr Johann Skuttle, for helping in various ways which included assistance in the collection of documentary materials, in arranging interviews with representatives of the Norwegian Government and other agencies concerned with petroleum development, for discussing the second chapter at a seminar in Bergen and for support towards the completion of the study; the Carnegie Endowment for International Peace and its Senior Research Associate, Dr Selig Harrison, for assistance in arranging interviews with representatives of US oil companies and government and other agencies concerned with petroleum; Senator Andres Sosa Pietri for assistance in arranging interviews in Venezuela; Dr Andrew R. Thompson, Chairman of the British Columbia Energy Commission and Dr Hasan S. Zakariya, UN Inter-Regional Adviser on

Petroleum Economics and Legislation not only for giving generously of their time but for providing valuable documentary material.

The ultimate responsibility for the contents and the judgements expressed on various matters, of course, rests exclusively with the author. Despite her protestations to remain anonymous, the help given by the author's wife, Hameeda Hossain, at all stages and in particular with reading of the proofs, cannot escape an expression of appreciation.

Sincere appreciation must be recorded of the valuable assistance rendered by the author's indefatigable Research Assistant, Tawfique Nawaz, of Wadham College, Oxford, who, immediately after taking his Honours Degree in Law, devoted himself with great dedication to work on the project. For typing a manuscript which was at times almost totally illegible thanks are due to Mrs Joan Peacock and to the dauntless band of secretaries of Nuffield College ably led by Miss Brotherhood and Mrs Yates.

Nuffield College   Kamal Hossain
Oxford, 20 April 1979

CHAPTER I

Petroleum Development: Historical Background and the Global Environment

"Oil's first century" dates from the middle of the nineteenth century.[1] From modest beginnings as a source for lighting lamps and fuelling stoves, oil, by the early nineteen-sixties had surpassed coal as the largest single source of energy. The growth of world oil production from 1 million tons in 1872 to over 2 billion tons in 1972 underlines the massive scale of petroleum development. Such development has taken place in a constantly changing global environment. The framework of law and policy governing petroleum development has in turn been subjected to continuing reappraisals and has undergone basic transformations.

The parameters of policy and the choice of legal instruments to secure policy objectives are largely influenced by the global environment. Before turning to a comparative examination of law and policy relating to petroleum development, it will be useful to delineate the historical background and the global environment in which such development has taken place.

HISTORICAL BACKGROUND

From the many well-documented accounts of the growth of the international petroleum industry,[2] certain salient features can be discerned. At the turn of this century, the United States and Russia accounted for more than 90 per cent of the world's oil. United States production exceeded its consumption, and a number of American firms exported petroleum products, amounting to about one-fourth of the total world consumption, outside the United States. Before 1900, American companies had no foreign oil production; by 1914, they had acquired petroleum interests only in Mexico and Romania.[3] The British and the Dutch had taken earlier initiatives in the development of foreign petroleum. The British Government had actively aided the grant of a concession in Iran to a British group in 1901; upon oil being struck in the concession area in 1908, the Anglo-Persian Oil Company (later

1

(Anglo-Iranian, now British Petroleum (BP)) was formed in 1909, in which the British Government proceeded to acquire a 50 per cent equity holding in 1914. The Royal Dutch Company was founded in 1890 to develop petroleum in the Dutch East Indies (Indonesia). It affiliated in 1907 with the Shell Transport and Trading Company, to form Royal Dutch Shell, owned to the extent of 60 per cent by Dutch interests and 40 per cent by British interests. Royal Dutch Shell was to emerge as the dominant petroleum producer in the Far East. The main oil-producing countries in 1913, and the quantities produced by them were as follows:[4]

Country	Quantity in tons
United States	33.0 million
Russia	8.6 million
Mexico	3.8 million
Romania	1.9 million
Dutch East Indies	1.6 million
Burma and India	1.1 million
Poland	1.1 million

The First World War established the importance of oil as a strategic raw material.[5] The war provided a powerful stimulus to the growth of the United States petroleum industry. At the end of the war, United States production amounted to nearly two-thirds of the world crude oil output and accounted for nearly 30 per cent of total oil consumption outside the United States. There was, however, growing concern about the pressure that such expansion imposed on United States petroleum reserves. The United States Government's response to this situation was to encourage foreign exploration.[6] American companies undertook commercial development in Venezuela in 1923. The United States Government exerted its influence to overcome the "exclusionary" policies of other governments in respect of prospective territories overseas, in particular in the Middle East. Such policies were reflected in the San Remo Agreement of 1920, and the Red-Line Agreement.[7] In 1928, a group of American companies, with the support of the State Department, acquired a 23.75 per cent interest in the Turkish Petroleum Company (later Iraq Petroleum Company). In the same year, Standard Oil of New Jersey (Jersey Standard) acquired concessions in the Dutch East Indies.

Major discoveries such as the large East Texas fields and expanding production resulted in large domestic surpluses in the United States in the early thirties. By 1933 United States exports of refined products had dropped by nearly half from their 1929 level, and the American export share of foreign consumption fell from 30 to 17.5 per cent.

During this period some of the major concessions in the Middle East were obtained by American companies. Standard Oil of California (Socal) obtained a concession in Bahrain in 1930. In 1933, in the face of strong opposition from the Iraq Petroleum Company, Socal obtained the concession in Saudi Arabia which was to form the basis of Aramco. Gulf Oil in the same year, and once again after overcoming opposition from a British-controlled company, the Anglo-Persian Oil Company, acquired a 50 per cent interest (the other 50 per cent being acquired by Anglo-Persian) in a concession in Kuwait.

The concessions granted in the Middle East up to 1945, the main characteristics of which are described below, "generally followed the same pattern and embodied the same standard conditions. Their similarity was equalled only by their simplicity":[8]

> The concession agreements defined the grant, the area and the duration of the concession, the payments to be made by the concessionaire and, mostly in general terms, the mutual rights and obligations of the parties. Where the law of the country required it, they were ratified by a law or a decree. In most cases, the agreements were described as concessions, though sometimes they were designated as conventions, contracts and leases.
>
> The grant made to the concessionaire was usually that of an exclusive right to search for, obtain, exploit, develop, render suitable for trade, carry away, export and sell petroleum and related substances.
>
> The area was very large and if it did not include the whole territory of the conceding State, it covered its largest part. The D'Arcy concession extended to the whole of the Persian Empire (Article 1) with the exception of five provinces (Article 6) and covered an area of 480,000 square miles. The area was reduced to 100,000 square miles in the new concession granted to AIOC in 1933 in replacement of the D'Arcy concession. IPC's concession (1925) was for the whole of Iraq east of the Tigris River, to the exclusion of the province of Basra and the "Transferred Territories". The concession originally granted by Saudi Arabia to Standard Oil Company of California, subsequently assigned to Aramco, covered an area of approximately 371,000 square miles and was further extended in 1939 to about 496,000 square miles. The concession granted to KOC covered the area of Kuwait. The concession for Bahrain (1934) was limited to 100,000 acres, but subsequently extended to include all of the Sheikh's present and future dominions. The area of the concession granted to the Anglo-Persian Oil Company in Qatar (1935) extended to the whole country.
>
> The duration of the concessions was quite long and usually ranged between 60 to 75 years. The original D'Arcy concession was made for 60 years. Concessions of the IPC group in Iraq were granted for 75 years. Aramco's original concession in Saudi Arabia was made for 60 years but was subsequently extended to 66 years. The concession

to KOC was for 75 years but was subsequently extended to 90 years. The concession in Bahrain (1934) was originally for 55 years, and subsequently extended to 90 years. The concession granted to Qatar to the Petroleum Development Company (1935) was for 75 years. The same period was stipulated in the agreement made in 1937 between the Sultan of Muscat and Oman and Petroleum Concessions Limited . . .

Judged by today's values, the financial terms of the earlier concessions seem modest. Although some concessions made provision for rents or tax commutation payments or fringe benefits, such as loans or the supply of petroleum . . . the principal feature of the oil concession was the royalty. The royalty was generally fixed at four shillings gold or three rupees per ton of crude oil.

These concessions were held by a small number of large international oil companies — Anglo-Persian (BP) in Iran, BP, Shell, Jersey Standard (and four American companies) in Iraq, Socal and Texaco in Saudi Arabia, BP and Gulf in Kuwait, and Socal in Bahrain. These companies and Mobil, which later acquired an interest in Aramco, are regarded as the seven major international companies and are referred to as "the majors".[9]

In Venezuela, the same companies (except BP) competed for concessions, and ultimately by 1937, 52 per cent of the total output was held by Jersey Standard, 40 per cent by Shell, and 7 per cent by Gulf.

In the Dutch East Indies, initially the Dutch colonial government accorded preferential treatment to Shell and other Dutch concerns, who were granted concessions of up to 75 years under which they paid a fixed price per acre and a percentage of the value of any oil produced.[10] Pressure from the United States, however, which included measures whereby Shell subsidiaries were denied leases on public lands in Utah, Wyoming and Oklahoma, resulted in American companies obtaining concessions there. By 1945, the oil industry in Indonesia was dominated by "the Big Three" companies — Shell, Stanvac (Jersey Standard and Standard Oil of New York) and Caltex (Socal and Texaco).

The main oil-producing countries in 1938, and the quantities produced by them were:[11]

Country	Quantity in tons
United States	161.9 million
Russia	29.3 million
Venezuela	27.7 million
Iran	10.2 million
Dutch East Indies	7.5 million
Mexico	5.5 million
Iraq	4.4 million

4

During the war, there was little further development in the Middle East or in other parts of Asia.

The international oil industry at the end of the Second World War was dominated by the majors. The shares of the total crude oil production (outside the US, Canada, USSR, Eastern Europe and China) held in 1950 by the majors were:[12]

Company	Quantity in tons	Per cent
Jersey Standard	50 million	25
British Petroleum	40 million	20
Royal Dutch/Shell	38.5 million	19
Gulf Oil	15 million	7
Texaco	12 million	6
Standard Oil (Calif.)	9 million	4
Mobil Oil	7 million	3
	171.5 million	85
Others	31 million	15
	202 million	100

The estimated reserves held by them in 1953 were as follows:[13]

Company	Amount (in billions of barrels)	Per cent
BP	28.3	35.7
Jersey Standard	11.2	14.2
Gulf	10.7	13.5
Shell	8.0	10.1
Texaco	5.4	6.9
Socal	5.2	6.5
Mobil	3.9	4.9

The dominant position of the major international oil companies gave them the power to regulate the rate of development of supply and to exercise strong control over prices, so long as they refrained from price competition among themselves. With large discoveries in Kuwait and Saudi Arabia, the majors were faced with the need for new marketing outlets. One solution adopted by them was for a company such as Socal, which had acquired a large source of supply in Saudi Arabia, to associate with Texaco, which had extensive Far Eastern and European markets. The majors basically aimed to produce the amount of petroleum which could be handled by their own marketing networks. At the end of the Second

5

World War, the majors constituted a "near-cartel — having agreed on market shares. Effects of the war and US anti-trust legislation eliminated formal agreements between them, but until the late 1950s — because of their organisation and control over supplies — they were able to work the industry the way they wanted it, with mutual understanding over pricing policies ensuring high profits for them."[14] They had also developed and controlled their own refining, transport and marketing — thereby achieving what is characterised as "vertical integration". It has been said that: "The combination of the vertical integration of these Companies with oligopoly in product markets provides the single most important key to the understanding of the economic policies of the industry, and of the past and present relations between oil Companies and governments of both the countries exporting and those importing crude oil or its products."[15]

The first decade after the Second World War was marked by a substantial rise in consumption, met to a substantial extent by the expanding output of new Middle Eastern sources of supply, a steady increase in the price of petroleum products and mounting profits accruing to the majors, in particular from their Middle Eastern operations.[16] This period has been referred to as the "ten golden years" for the major international oil companies.[17] Up to 1958, the rate of profit earned in foreign investments in petroleum was conspicuously higher than the rates earned on foreign investments in manufacturing, mining, and other industries. In fact the rate for oil averaged more than thrice the rate on investments in the mining industry, which also carried relatively high risks.[18]

World consumption increased from 250 million tons in 1941 to 950 million tons in 1955, that is at a rate, compounded, of 7 per cent per annum.[19] The main sources from which this increasing demand was met is seen from the list of the main producing countries in 1950 and in 1957:

1950[20]

Country	Quantity in tons
United States	295 million
Soviet Union	213 million
Venezuela	75 million
Iran	33 million
Saudi Arabia	29 million
Kuwait	17 million
Iraq	6.8 million
Indonesia	4 million

1957

Country	Quantity in tons
United States	360 million
Soviet Union	175 million
Venezuela	142 million
Kuwait	57 million
Saudi Arabia	49.5 million
Iran	36 million
Iraq	22.6 million
Indonesia	21 million

Developments in the first post-war decade accelerated changes in the course of the second decade. These changes "were largely the result of three mutually reinforcing developments: (1) the increased competition for concessions to explore for and produce crude oil, including the entry of newcomers and the rise of new sources of crude, which changed the terms on which concessions could be obtained; (2) increased competition in crude and product markets, which directly and indirectly affected prices and brought changes in the policies of some importing countries; (3) the increased importance of government policies".[21]

The high level of the profits earned by the majors, and the increasing demand for petroleum, provided the impetus for large-scale exploration. The majors as a result of the large Middle Eastern discoveries held a vast "surplus" − "the conservatively estimated addition to Middle Eastern oil reserves alone between 1947 to 1962 was 187 billion barrels (27 billion tons)".[22] The situation of the majors during this period has been described thus:

> Crude oil producers were few, and entry was extremely slow. No producer would sell off any part of his reserves although their present value was nil . . . He preferred to keep an inventory which might be 50 to 150 times sales, in the volumes of oil out of new competitors' hands. Entry was therefore difficult and unpromising, and although some succeeded, none of them did so on a scale to rival the older established producers. The huge resources of the original Persian Gulf concessions were effectively locked up, and only by the late 1960's was there a comparable rival, in Libya. In Venezuela, concessions were granted only once in the entire post-war period, in 1956. Although some good discoveries were made, no great volume of oil was available to compete with the Persian Gulf producers, whose overlapping joint ventures allowed the co-ordination of output and hampered individualistic sales and production policies. Vertical integration with refining-marketing precluded any class of large and well-informed

7

buyers who would have helped induce competitive conduct. It also permitted mutual surveillance, for refining provided a check on any given company's producer plans.[23]

In 1945, only six US companies, other than the five majors, had active exploration interests — Atlantic, Blue Goose (Ganso Azul), Cities Service, Philips, Richfield, and Sinclair. In 1953, twenty-eight American firms other than the five largest companies possessed foreign exploration rights. Ten had ventures in the Eastern Hemisphere and twenty in Latin America. Only four had interests in more than one country.[24] Before the formation of the Iranian Consortium in 1954, consisting of nine "independent" American companies, there were a handful of independents in the Middle East, the only two of importance being Aminoil and Pacific Western Oil (later Getty). It was in the second post-war decade that there was a massive entry of "new-comers", including "independents" into the field of petroleum exploration. During the period, 1953 to 1972, more than three hundred private companies either entered the foreign oil industry or significantly expanded their participation in it.

The new entrants were extraordinarily diverse in character, involving enterprises of many sizes and nationalities. Some entries were made by petroleum companies that had previously confined their operations to their own national markets. In other cases, companies which had been engaged in one or two markets or stages of industry outside their home countries integrated their foreign operations forwards towards markets, backward towards production, or geographically. In still other instances, entries were made by companies newly organised for the purpose, or by firms previously engaged in natural gas, chemical steel, automotive or other industries . . . Many government-owned or government-sponsored companies entered the world's oil markets, usually in competition with private companies, but sometimes with national monopoly rights or special government preferences. The entry of the giant Soviet Oil Trust was a signal event of the late 1950's.

Among the 350 different firms that entered the foreign oil industry between 1953 and 1972, it is convenient to classify the *principal* ones into five groups: fifteen large US companies, ten large US natural gas, chemical and other non-oil companies, twenty-five foreign private corporations, and fifteen foreign government oil companies.

The fifteen large US oil companies which had by 1972 become vigorous competitors of the established firms included Amerada Hess, Aminoil, Atlantic-Richfield, Cities Service, Continental, Getty, Iricon, Occidental, Pan-American, Phillips, Sinclair, Ohio Oil Company (now Marathon), Sun, Superior and Union . . . By the end of 1972, each of them had found, and a dozen of them were producing, significant quantities of oil (100,000 barrels per day or more). The majority of them were operating in six to twelve countries. Most had substantial foreign refining or marketing interests. The assets of eleven of the group totalled $20 billion, or an average of more than $1.9 billion each. *Twenty medium size US oil companies* also went overseas on a considerable scale. These included Ambassador, Ashland, Clark, Daho-American, Delhi Taylor, Franco Wyoming, General American, General Exploration, Hunt, Kern County Land (later in Tenneco) Kerr-McGee, Mecom, Murphy, Natomas, Standard (Ohio), Pan-Coastal. Pauley, Pure (later merged into Union), Sunray and Texas Gulf Producing . . . [By 1962] fourteen of them had struck oil in commercial quantities.

Ten large US natural gas, chemical and steel companies also entered the field of foreign oil seeking profitable diversification. They included El Paso, Tennessee, Texas Eastern, United, and Western and Colorado (natural gas companies); Allied Chemical, Monsanto, W. R. Grace, and United Carbon (chemical companies); and Detroit (steel company). Up to the end of 1962, six of them had found oil . . .

Twenty-five foreign private firms entered the international oil industry on a major scale. The group included, from Australia, Ampol (and over sixty different Australian firms, often in joint venture with United States, British or Canadian firms); from Belgium, Petrofina; from England, Ultramar; from France, Antar; from Israel, Paz; from Italy, Ansonia Mineraria and Montecatini; from Japan, the Arabian Oil Company (AOC), Daikyo, Idemitsu, Maruzen, Nippon Mining, and North Sumatra; from Spain, CEPSA; from Sweden, IK and Nynas; from Switzerland, Avia and Raffineries du Rhone; and from West Germany, Aral, DEA, Elwerath, Frusia, Gelsenberg, Preussag, and Wintershall . . . Some were newly-formed companies, which (like AOC) had, by 1962, attained striking success. Some German companies sought to integrate established coal operations horizontally into the refining and marketing of oil, and refining operations backwards to production. Some (like Ansonia Mineraria and Montecatini) were giant enterprises seeking product diversification. Their ample financial resources made them, as well as many of the others, formidable potential or actual competitors.

Fifteen foreign government companies began or significantly extended their oil operations between 1953 and 1972 . . . The leading companies were, from Argentina, YPF; from Brazil, Petrobras; from Egypt, EGPC; from France, CFP; from India, IOC; from Indonesia, Pertamina; from Iran, NIOC; from Iraq, INOC; from Italy, ENI; from Kuwait, KNPC; from Mexico, Pemex; from Saudi Arabia, Petromin; from Spain, Campsa; from Turkey, TPAO; and from Venezuela, CVP . . . [25]

Between 1953 and 1972 there was a massive expansion of petroleum exploration. In 1953 the total area in respect of which exploration rights were held outside the country of the operating entity concerned, was 520,000 square miles, held in some twenty-five foreign countries by a total of seventy-seven private entities, and 495,000 square miles held by four government companies in fifteen other countries; in 1972, more than 330 companies (other than the majors) held exploration rights to 6.8 million square miles, located in 122 areas of the world. In 1953 no private oil company, other than the seven majors, had as much as 200 million barrels of proven foreign crude oil reserves, and among the government companies only one had reserves of this figure. By 1972 no less than thirteen other companies each owned more than 2 billion barrels of oil in the ground. Collectively, all the new entrants (apart from the Soviet Union) owned about 112 billion barrels of proven reserves, amounting to almost one-fourth of the total reserves, outside the socialist world. These reserves, produced at the conservative rate of 5 per cent per year, could maintain a production of 15 billion barrels per day, or 58 per cent of the 1972 rate of world consumption (excluding the socialist world), for the next 20 years. With regard to production, in 1953, no company − other than the majors − had daily foreign oil production of 200,000 barrels or more. By the end of 1972, thirteen of the new companies had passed this level, and collectively, the newcomers had attained a daily output of about 5.2 million barrels, or more than one-sixth of the world consumption (excluding the socialist world).[26]

Another significant development during this period was the expansion of Soviet petroleum exports, outside the socialist world, which increased from 3 million tons in 1955 to over 50 million tons in 1972.

Oil prices had risen by about 25 per cent during 1949-57. The re-opening of the Suez Canal in 1957 marked the beginning of a period of price decline. New crude oil producers, without refining facilities or marketing networks of their own, sought outlets. The oil producers competed with the new entrants to find new markets or to retain their established ones, leading to offers of crude oil at substantial discounts, below the "posted prices" of the majors.[27] The majors began to give discounts to non-affiliated buyers on long-term contracts; Soviet oil exports offered at prices below "world prices" found willing buyers. The United States imposed an import quota in 1959, thus reducing access to the US market to American companies who were thus left to find alternative markets in Europe and elsewhere − at lower prices. In this situation, the majors proceeded to reduce the posted price in 1959. A further reduction in 1960 provoked the oil-producing countries, alarmed by the downward tendency in oil prices, to form the Organisation of the Petroleum Exporting Countries (OPEC).[28] It also led to a reappraisal by those countries of the arrangements under which petroleum development and production was being carried on. This reappraisal was to result in raising the level of awareness about public policy issues relating to petroleum and to the emergence of new frameworks for petroleum development in the established petroleum-producing countries.

During the 1960s, while an accelerating inflation continued to raise the price levels in most of the industrialised world (in the United States, the consumer price index rose 23 per cent between 1965–70), petroleum prices remained virtually constant. This meant a steady worsening of the terms of trade for the OPEC countries, who were trading their low-priced oil for increasingly expensive industrial products. In the same decade, there was substantial rise in petroleum consumption in the industrialised countries, and an increase in imports of crude oil from the OPEC countries.

The growing demand for petroleum in the industrialised countries led to a steady increase in imports. The United States by 1972 was importing nearly 30 per cent of its crude oil requirements — the main sources for such imports was Latin America which accounted for about 25 per cent; about 5 per cent was imported from the Middle East and North Africa. The Middle East and North Africa accounted, in 1972, for 78.6 per cent of Japan's oil imports, and 79.5 per cent of Western Europe's.[29]

The bargaining situation between the OPEC members and industrialised countries thus underwent a fundamental change. The increased bargaining power was first exercised by Libya, which had acquired added importance as a source of supply for Western Europe as a result of the closure of the Suez Canal in 1967. This advantage had been enhanced with the shut-down in May 1970 of the Trans-Arabian Pipeline which transported 500,000 barrels of crude oil daily from Saudi Arabia to the Mediterranean. The new revolutionary government in Libya in September 1970 successfully pressed for a higher price from one of its concessionaires, Occidental Petroleum, and followed with similar demands against the other concessionaires. A cut-back of crude oil production was effected to exert pressure, and a total stoppage of production was threatened, in order to secure compliance with its demands.[30] Inevitably this encouraged other OPEC countries to press for price increases. Collectively, OPEC in

December 1970 called for negotiations with the oil companies on its demands which included establishment of a 55 per cent income tax rate, an increase of posted prices, elimination of special discounts on sales of crude oil and disparities in prices among OPEC members. The Teheran Agreement of 1971 was based on a substantial acceptance of these demands. Further agreements were reached in 1972 and 1973 raising the

11

price first by 20 cents per barrel, and then by 15 cents, to offset the effect of the devaluation of the dollar. In September 1973 OPEC raised the issue of renegotiation of the Teheran Agreement on the basis that the current rate of inflation exceeded 2.5 per cent upon which that agreement had been premised. The negotiations were barely under way when the October war erupted and in its wake OPEC members unilaterally increased the posted price. The Arab oil producers led by Saudi Arabia in a demonstration of unity, which was to sustain their capacity to maintain such price increases, announced a policy of graduated reductions of oil production and a selective embargo.

These events brought about what has been characterised as "the world oil revolution", which defined new parameters within which policies relating to petroleum would henceforward be made.

THE GLOBAL ENVIRONMENT

It is evident from the historical background of the industry that the global environment for petroleum development is determined by the structure of the industry, and the inter-action of the policies aed actions of: (a) the international oil companies and their home governments; (b) the major oil-producing countries; and (c) policies of oil-importing countries (both developed and developing).

The economic characteristics of the industry are: the oligopolistic structure of the multinational oil companies, the formal or informal vertical integration of the industry, the high growth rate of demand, the non-competitiveness of alternative sources of energy and the small number of major exporting countries.[31]

Up until 1945 (and for nearly two decades thereafter) the global environment worked to the advantage of the major multinationals. The structure of the industry, the substantial resources of the companies, which included capital, technology, and control over markets and sources of supply, and the power wielded by their home governments enabled them to acquire, and to maintain, dominance in the international petroleum system and gave them an overwhelmingly stronger bargaining position in relation to the governments with which they had dealings.

The companies could plan and act globally — establish priorities for exploration and development among different regions and different countries, determine production quotas for each of the producing countries, and maintain a world price level. They had been able to obtain valuable concessions on specially advantageous terms. These were advantageous not only in financial terms, which required payment of a fixed royalty per ton, but because they secured maximum freedom of action for the companies. The companies were left free to control every aspect of the operations, to determine the pace and extent of exploration, the rate of development, the rate and levels of production, and to fix the price

The end of the war saw the emergence in the established oil-producing countries of nationalist feelings which expressed dissatisfaction with the existing concessionary system, and demanded changes in it. The urge to assert "national sovereignty" over natural resources initially focused on the high profits covered by the companies and the relatively low returns to the governments. Thus, a larger "government take" was the first of the objectives aimed at. Venezuela led the way by enactment on 12 November 1948 of an income-tax law which aimed to secure a "fifty-fifty" distribution of profits between the government and the oil companies — a measure characterised as "a concept of taxation that was then revolutionary in the oil industry . . ."[32] The companies accepted the "fifty-fifty" profit sharing formula in Saudi Arabia, and the decree of 27 December 1950 enacted this into law. The formula was adopted in Kuwait on 29 December 1951 and in Iraq on 3 February 1952.[33] This only began the process by which the governments would claim an increasingly larger share of the "economic rent" generated by "their" natural resources.

While the companies sought to contain the pressures for change by accommodating the demand for increasing financial returns, they were far less amenable with regard to demands calling for other alterations in the concessionary system, in particular those which could erode their control over operations or over markets.

In Iran, the demand for "Iranisation" of the industry in 10 years and for access to the company's books, could not be assuaged by the offer of a "fifty-fifty" profit-sharing formula. The company's unwillingness to consider these demands precipitated the nationalisation of AIOC.[34]

· The relative strength of the company and the relative weakness of the government of a major oil-producing country in the environment of the fifties is evident from the retaliatory action which the company was able to take, by way of an embargo. AIOC enforced an effective embargo against the production of the nationalised enterprise, with the

13

co-operation of the other majors who between them controlled 90 per cent of the marketing outlets in the Eastern Hemisphere: the annual production in Iran fell from 241 million barrels in 1950 to 9 million barrels in 1953. Normal production levels could be restored only after operations were entrusted to a Consortium, consisting of AIOC, Shell and a number of American companies. The government admitted, while seeking approval from the Majlis of the Consortium agreement, that it was "not an ideal solution", but no better solution could be attained, for as the government spokesman said: ". . . we do not have the means to compete in international markets because we do not possess a marketing organisation."[35]

The practice of "sitting on" a concession, namely retention by concessionaires of concession areas, without devoting any effort towards exploration and development in those areas, was another feature of the concessionary system which was assailed. Majors who controlled numerous sources of supply could justify, in terms of rational company behaviour, inaction in respect of portions of concession areas held by them, so long as they had adequate, and even surplus, capacity available to them in other areas to meet the needs of their marketing outlets. At the same time they were unwilling to relinquish these areas, so as to preclude the possibility of these areas being acquired by their potential competitors. The IPC concessions in Iraq provided some clear instances of this practice:[36]

> Of the many concession areas exclusively preempted IPC, none have been rapidly developed . . . Qatar is another illustration of "sitting on" a concession . . . Anglo-Persian and Shell . . . were not anxious to develop more production in the Persian Gulf because of the effect it would have on production in Iran . . . (the general manager of IPC wrote:) ". . . we have been steadily complying with the letter of the Mining Law *by drilling shallow holes on locations where there was no danger of our striking oil.*"

In the negotiations between the Iraqi Government and the companies in the fifties, relinquishment of areas was an important issue. The failure to reach an agreement on the issue was followed by a law issued on 11 December 1961 by which the government unilaterally resumed 99.5 per cent of the concession areas, allowing the company to retain only 1937.75 square kilometres.[37]

The power reserved to the companies to determine rates of development and levels of production led to the grievance on the part of some governments that development of their oilfields was being held back in order to serve the commercial interests of the companies. The Prime Minister of Iran averred:[38]

The growth and development of Iranian society must not be subjected to the fluctuations caused in production and export of oil as a result of unilateral decisions taken by foreign boards of directors behind closed doors and outside Iran . . .

And, the Shah emphasised the importance attached to this issue thus:[39]

It comes down to the point when we have to decide whether the oil companies which are, after all, only commercial undertakings, have the right to refuse to extract from our soil our own resources. This cannot be accepted. And what if no settlement of this difference is possible? We may in that case have to act unilaterally – that is, to depart from our agreement with the Consortium.

Thus, in the established oil-producing countries, the concessionary system was under pressure; it was increasingly felt that the unfettered discretion that the companies enjoyed in respect of operations and key decisions involving pricing and production acted in numerous ways to the detriment of the state. Among the issues which have been catalogued as sources of grievance among the Middle Eastern oil producers are:[40]

(a) long duration of concessions;
(b) absence of relinquishment provision or where such provision existed, failure to comply with its requirement;
(c) suspension of the right to tax;
(d) discretion given to concessionaire to determine pace of exploration, and to decide what areas should be developed and for how long areas should lie idle;
(e) managerial decisions being exclusively in the hands of foreigners;
(f) flaring of natural gas, and failure to re-inject the gas back into reservoirs to provide for greater ultimate recovery;
(g) arbitrary pricing policies;
(h) adoption of accounting methods and procedures which had the effect of reducing "government take".

While efforts were exerted to revise existing arrangements through negotiations, the end of the fifties and the early sixties witnessed developments which ushered in major transformations in the global environment for petroleum development. The urge in the producing countries to exercise control over the development of their petroleum resources and to reduce the market power of the majors, as well as dependence on them, led them (a) to establish national oil companies (b) to grant exploration rights to new entrants in the field under new types of petroleum development arrangements. These new "entrants" included "independent" oil companies, that is, privately owned foreign companies other than the majors, and state-owned oil companies established by some of the major consuming countries. Exploration activities were extended to many new areas; some of these were to become important sources of supply, in particular in North and West Africa. And it was in 1960 that OPEC was established by the major exporting countries as an organisation through which they could concert their efforts

> to study and formulate a system to ensure the stabilization of prices by, among other means, the regulation of production, with due regard to the interests of the producing and of the consuming nations and to the necessity of securing a steady income to the producing countries, an efficient economic and regular supply of this source of energy to consuming nations, and a fair return on their capital to those investing in the petroleum industry.[41]

The Arab oil-producing countries formed an organisation of their own, the Organisation of Arab Petroleum Producing Countries (OAPEC) in 1968. The state oil entities in Latin America formed a regional organisation "Asistencia Reciproca Petrolera Estatal Latinoamericana" in 1965.

While state-owned oil companies had played a part in some of the Latin American countries since the twenties,[42] national oil companies designed to assume a central role in national petroleum development emerged in the wake of nationalisation of the industry, first in Mexico in 1938, where PEMEX (Petroleos Mexicanos) was formed, and then in Iran in 1951, when the NIOC (National Iranian Oil Company) was established. The spate of national oil companies which came into existence in the fifties and sixties were a response conditioned by nationalist feelings to what was perceived as excessive dependence on the major international oil companies. This dependence, so far as producing countries were concerned, resulted from lack of technical and managerial capabilities and marketing outlets; in the case of consumer countries, from lack of alternative sources of supply, which led them to seek "security of supply" controlled by their own national agencies, which could also be expected to meet national requirements at a lower cost.[43]

In Iran the Petroleum Act of 1957 designated NIOC as the instrument for accomplishing its objectives, which were the exploration for and exploitation of oil *as rapidly as possible* in all areas outside those assigned to Consortium, and development of refining, transportation and sale of oil throughout Iran and abroad. The Corporacion Venezolana del Petroleo (CVP) was established in April 1960: its *raison d'être* has been described thus:

> Among the reasons given for its establishment were the financial benefits expected to accrue to the country: the government would

receive 100 per cent of CVP's profit in contrast to the mere 60 or 70 per cent of private oil company profits obtained through taxation. CVP was also to provide the government with a standard of comparison by which to judge the performance and contentions of the private companies, and it was to offer a training ground for technical personnel who would be able to run the oil industry with the aid of service contracts at the expiration of the current concessions.[44]

The Kuwait National Petroleum Company was incorporated in October 1960. Saudi Arabia established its General Petroleum and Mineral Organisation (Petromin) in 1962. Algeria constituted its national oil company, the Société national pour le transportation et la commercialisation des hydrocarbures (Sonatrach) in 1963. The Iraq National Oil Company (INOC) was established in 1964. Indonesia's Pertamina emerged in 1968 as did the Libyan Petroleum Company (later re-named the Libyan National Oil Corporation). The Nigerian National Oil Corporation was formed as was the Abu Dhabi National Oil Company in 1971.[45]

Among the consumer countries, Italy's Ente Nazionale Idrocarburi (ENI) formed in 1953, and France's Entreprise des Recherches et d'Activités Petrolières (ERAP) constituted in 1965, took important initiatives to undertake overseas exploration programmes and to secure for their countries new sources of supply. Japan established in 1967 the Japan Petroleum Development Corporation, a state-controlled agency to coordinate and promote exploration and development by Japanese companies, by extending financial, material and technical assistance to them.

The interest of producing countries and that of some of the state-owned oil companies of the consumer countries and of the "independent" oil companies coincided to the extent that both were interested in undertaking petroleum development *independently* from the major oil companies. Producing countries welcomed these new entrants into their countries to engage in petroleum development under new types of arrangements. In these new arrangements, conscious attempts were made to devise provisions which would remove the principal sources of grievance felt by the host governments with the old concessionary system.

Thus, in 1957 ENI pioneered "joint venture" arrangements in Egypt and in Iran.[46] Iran followed this model in entering into arrangements with "independents" — the Pan American Petroleum Corporation, a subsidiary of Standard Oil of Indiana and Sapphire International Petroleum Limited of Canada. Venezuela declared a policy of "no more concessions" in 1958, and CVP, its national oil company, devised "service contracts" under which it could engage a foreign oil company to carry out petroleum

17

exploration and development in consideration for payment for services rendered, under CVP's supervision and without conceding to the foreign company any property rights or management prerogatives.[47] The first such service contract was concluded in September 1962. The "service contract" type arrangement was introduced in the Middle East by ERAP, which entered into such an agreement with INOC in Iraq in 1968.

Indonesia enacted a law, Law No. 44 of 1960, which provided that "petroleum operations" could henceforward only be undertaken by the state and exclusively carried out by state enterprises. Existing operators were required to terminate their operations under the old concessions, but were eligible to enter into "contracts of work" compatible with the new law. The engagement of an "independent" the Pan American Oil Corporation on 15 June 1962, induced the majors who where operating in Indonesia to overcome their initial reluctance to enter into similar "contracts of work", which they did on 25 September 1963.[48] These contracts were the precursors of the "production-sharing contracts", an Indonesian innovation — the first of which was concluded with a new company, formed by a group of American operators — the Independent Indonesian American Petroleum Company (IIAPO) on 18 August 1966.[49]

Algeria negotiated with France to evolve an inter-governmental "Co-operative Association" for the joint exploration and production of hydro-carbons under the Franco-Algerian Accord of 29 July 1965.[50] It also adopted an improved version of the "joint venture" model in which the host government was given a 51 per cent interest; the first such agreement was entered into between Sonatrach and Getty Petroleum Company on 19 October 1968. In 1969, Iraq entered into an agreement with the Soviet Union under which it would be provided with machinery, materials and technical assistance necessary to develop INOC into a fully integrated national oil company.

The assertion of sovereignty over natural resources, reflected in national petroleum policies, was supported by the doctrine of "Permanent Sovereignty over Natural Resources", affirmed by the United Nations General Assembly, first in its Resolution No. 1803 (XVII) in 1960 and then in a more comprehensive form in Resolution No. 2518 (XXI) in 1966. The latter resolution while reaffirming "the inalienable right of all countries to exercise permanent sovereignty over their natural resources in the interest of their national development": and that "the United Nations should undertake a maximum concerted effort to channel its activities so as to enable all countries to exercise that right fully" goes on to state "that such an effort should help in achieving the maximum possible development of the natural resources of the developing countries and *in strengthening their ability to undertake this development themselves, . . .*[51] OPEC in its Declaratory Statement of Petroleum Policy in 1968 reflected the same orientation and spelt out the principles underlying the new types of arrangements for petroleum development, thus:[52]

> Member Governments shall endeavour, as far as feasible, to explore for and develop their hydrocarbon resources directly. The capital, specialists and the promotion of marketing outlets required for such direct development may be complemented when necessary from alternate sources on a commercial basis.
>
> However, when a Member Government is not capable of developing its hydrocarbon resources directly, it may enter into contracts of various types, to be defined in its legislation but subject to the present principles, with outside operators for a reasonable remuneration, taking into account the degree of risk involved. Under such arrangements, the Government shall seek to retain the greatest measure possible of participation in and control over operations . . .
>
> Where provision for governmental participation in the ownership of the concession-holding company under any of the petroleum contracts has not been made, the Government may acquire a reasonable participation on the grounds of the principle of changing circumstances . . .

The developments of the sixties were to alter the relative bargaining positions of the international oil companies and the producing countries. The impact of new entrants led to a substantial expansion in output, and "excess productive capacity". "Indeed, it has been argued that during the 1960s the rising pluralism and heterogeneity of the industry led to *excessive* competition of a destabilizing kind . . . The industry became composed of so many firms having such diverse interests as to be incapable of offering an adequate counterpoise to the rising power of the oil nations."[53]

OPEC's record in the sixties is one of modest achievements. Its members were able to demonstrate "the potential power of a group of nations against the international companies, which had hitherto been able to 'play off' any one of the countries against the others".[54] They were able collectively to work out and secure the agreement of the companies to an alteration of the practice of deducting "royalties" from income-tax liabilities which reduced the revenue of the government. The 1964 agreement provided that "royalties" would be treated as an "expense" rather than be deducted from the income tax payable to the government. In 1968, they were able to secure an agreement which would phase out the discount allowance on the posted price.[54] OPEC had not, however, succeeded in getting its members to agree to the pro-rationing or programmed regulation of output. 1970 is said to mark the turning point, when OPEC was able at its Caracas meeting to formalise a collective demand for increased royalties and taxes, and what is more significant to secure a substantial increase in oil prices. This "demonstrated that the balance of power in the oil world was moving away from the oil companies and in favour of the nations with oil resources".[55]

The combination of circumstances in which OPEC could achieve this result have been described thus:[56]

> The first signs of upheaval appeared at Rotterdam, the oil port where prices for additional tonnages needed by refiners on a large scale are decided. Heavy fuel oil was still selling at 9 dollars a ton in September 1969 but rose to 24 dollars by the end of 1970 . . . There were many reasons for this price rise.
>
> There was an insatiable demand from Japanese industry for energy sources. Because of this the Japanese interests were taking every opportunity to buy in all the main markets of the world, which put the price up. The efficacy of the fight against pollution in the United States obliged middlemen . . . to buy in Europe the North African petroleum products with lower sulphur content than those on the domestic market in the USA. Consumption was increasing so fast, having doubled in nine years that it was outdistancing the estimates of the experts, especially in cold weather – because of the rise in consumption of domestic oil. Supply from the Mediterranean region had fallen short of the companies' expectations. Libya had unilaterally imposed restrictions of [sic] production in 1970. Tapline, carrying Saudi Arabian crude to the Mediterranean, had been closed for nine months, because of an "accident" to the pipeline in Syrian territory. The gap between supply and demand caused enterprises to turn more definitely to the oilfields in the Middle East, but there was insufficient tanker capacity available to cope with this sudden rush of trade. Hence there was a normal rise in oil prices and an abnormal speculative rise in freight rates from the Persian Gulf to Europe . . . Because of the world situation the advantage suddenly tipped in favour of the producer countries banded together in OPEC. They could not miss such an obvious opportunity.

Having attained a significant improvement in financial terms, in the Teheran Agreement of 1971, OPEC turned its attention towards increasing control over operations. By its Resolution No. XXIV. 135, adopted in July 1971, it urged member countries to take immediate steps towards the actual implementation of the principle of *participation* in existing oil concessions. Protracted negotiations with the companies yielded the "Participation Agreement", which became effective on 1 January 1973, and provided for an initial participation by host governments of

25 per cent in each concession, to be progressively increased to 51 per cent by 1981.[57]

The pace of developments, however, could not be confined within the evolutionary framework negotiated by the parties. The October war in the Middle East in October 1973 acted as a catalyst to precipitate a totally new situation.[58] OPEC members unilaterally decided to raise prices in October 1973, so that between 1973 to 1974, there was a quadrupling of oil prices. The posted price per barrel of $2.59 prevailing in early 1973 was raised by Saudi Arabia to $11.65 per barrel in January 1974. Member countries also took decisions to cut back production and to impose embargoes, thus assuming the power to control production. As the result of a series of nationalisation measures, and an acceleration of the participation process in disregard of the time-table earlier agreed upon, most of the OPEC members had by 1975 acquired majority, if not total ownership of oil reserves and production facilities.[59] Thus, control over pricing and production shifted from the companies to the producer governments, as did control over reserves and production facilities. Between 1970 to 1975, OPEC would appear to have fulfilled most of its important objectives:

> Not only was the posted price of oil restored to its pre-1960 level, but it has surpassed that by leaps and bounds. The sole authority of its members to fix the level of prices and the rate of production periodically is now well established and unchallenged. The concession regime is a thing of the past in some countries and is in the last stage of liquidation in others. Member states are recouping most, if not all, the so-called "economic rent" of their petroleum, leaving the foreign operators with only what they are entitled to: a reasonable and equitable margin of return on their investments as far as production is concerned. The national oil companies are progressively consolidating their positions in the world market and consummating deals of increasing importance.[60]

It is this totally changed setting that has led governments to accord the highest priority to the issues of energy policy, to take their bearings in the entirely new global environment, and to reappraise their own existing policies and arrangements for development of petroleum and other sources of energy.

Given the magnitude of the impact of the new global energy situation, governments have been called upon not only to formulate a "national energy policy" and a strategy for meeting national energy needs, but to assume a greater and more active role in energy development. The present situation provides a strong impetus to exploration for petroleum, as it does to the development of new sources of energy. For every country which is a net importer, and this is the position of most of the countries (*vide*

21

Table A appended to this chapter), a very high priority is now accorded to undertaking further exploration and development. Wherever there is a possibility that indigenous petroleum occurrences might exist, there is now a strong incentive to search for them at a much higher cost than was previously considered feasible. It is also regarded as advantageous to seek new sources of supply which are not controlled by OPEC countries. It has been argued that: "As long as Middle East resources seemed infinite and production from the area seemed destined to go on increasing at dramatic rates indefinitely, there was always a cost and supply disincentive for exploration in other parts of the world outside America. Now that there is every reason to find alternatives to the Middle East and every price incentive to make the effort worthwhile, the rate of exploration, the areas in which it is carried out and the pace of development of finds is bound to go up enormously"[61] and that ". . . with OPEC oil at around $8 per barrel, it would be profitable for the oil companies to develop oil resources of the deep jungles, the arctic wastes and the ocean beds of the continental shelves."[62] With OPEC oil at around $12 per barrel, the argument for rapid expansion of exploration in new areas is further strengthened.

There is a perceptible increase of interest in exploration of new areas, and many countries, which have hitherto had little or no experience of searching for oil are being drawn into the field of petroleum exploration. They are thus called upon to formulate policies and establish a framework of law and policy for undertaking this task. Non-OPEC producing countries, like Britain, Canada and Norway, have felt the need for a policy review in order to define their responses to the new situation, leading in a number of cases to the revision of existing policies, to alterations in the arrangements under which petroleum exploration was being carried on, and to a good deal of new legislation in the field of petroleum development. It is to an examination of the policy issues and options which have been under consideration by governments, and the decisions and choices, which have been made by them, that we shall turn in the next chapter.

Table A

PRODUCTION, IMPORTS AND EXPORTS OF CRUDE PETROLEUM
(Country-wise, 1974 — in million metric tons)

Country	Production	Imports	Exports
AFRICA			
1. Algeria	48.660		41.861
2. Libya	73.364		104.861
3. Morocco	0.024	2.262	
4. Sudan			0.700
5. Tunisia	4.139	0.843	3.726
6. Egypt	7.472	2.000	2.320
7. Congo	2.455		2.450
8. Gabon	10.202		9.070
9. French Equatorial Africa			
10. Angola	8.700		7.800
11. Zaïre		0.678	
12. Ethiopia		0.680	
13. Ghana		0.879	
14. Ivory Coast		1.418	
15. Kenya		2.808	
16. Liberia		0.595	
17. Madagascar		0.680	
18. Mozambique		0.255	
19. Nigeria	111.578		108.014
20. Senegal		0.673	
21. Sierra Leone		0.310	
22. Southern Rhodesia			
23. Tanzania		1.638	
24. Zambia		0.840	
NORTH AMERICA			
25. Canada	82.531	40.590	40.331
26. USA	432.794	172.573	0.145
OTHER AMERICA			
27. Argentina	21.139	2.380	0.067
28. Bolivia	2.112		1.374
29. Brazil	8.442	32.731	1.654
30. Chile	1.311	3.920	
31. Ecuador	8.999	0.850	8.342
32. Paraguay		0.229	
33. Peru	3.756	1.510	0.190
34. Uruguay		1.700	
CENTRAL AMERICA			
35. Costa Rica		0.400	
36. El Salvador		0.610	
37. Guatemala		9.050	
38. Hondoras		0.740	
39. Nicaragua		0.560	
40. Antigua		0.420	
41. Bahamas		10.000	
42. Barbados		0.145	

Table A *continued*

Country	Production	Imports	Exports
43. Colombia	8.225		0.068
44. Cuba	0.140	5.350	
45. Dominican Republic		0.950	
46. Jamaica		2.220	
47. Martinique		0.516	
48. Mexico	35.550	14.660	6.718
49. Neth. Antilles		40.500	0.570
50. Panama		3.510	
51. Puerto Rico		16.450	
52. Trinidad	10.550	14.660	6.718
53. US Virgin Islands		24.400	
54. Venezuela	155.803		92.802

MIDDLE EAST

55. Israel	6.040	0.860	
56. Bahrain	3.363	9.239	
57. Cyprus		0.510	
58. Iran	300.852		268.273
59. Iraq	96.940		88.970
60. Jordan		0.773	
61. Kuwait	128.101		110.848
62. Lebanon		2.500	
63. Oman	14.488		14.480
64. Qatar	25.059		24.638
65. Saudi Arabia	421.397	0.500	392.761
66. Democratic Yemen		2.700	
67. Syria	6.426	1.726	6.160
68. UAE	82.071		81.071
69. Turkey	3.110	9.961	

FAR EAST

70. Japan	0.672	237.579	0.003
71. Okinawa		3.285	
72. Bangladesh		0.575	
73. Brunei	9.284		9.225
74. Burma	0.888	0.250	
75. Cambodia			
76. Sri Lanka		1.800	
77. India	7.490	14.629	
78. Indonesia	67.979		50.829
79. Korea		15.508	
80. W. Malaysia		2.850	0.100
81. Sarawak	3.844	0.692	3.025
82. Pakistan	0.432	3.161	
83. Philippines		8.485	
84. Singapore		23.059	
85. Thailand	0.010	7.000	
86. Malaysia/Singapore			

Table A *continued*

Country	Production	Imports	Exports
CENTRALLY PLANNED ASIA			
87. China	65.000		4.000
88. Mongolia		0.004	
WESTERN EUROPE			
89. Belgium		30.575	0.102
90. Denmark	0.086	9.283	
91. France	1.080	130.203	
92. West Germany	6.191	104.459	
93. Ireland		2.727	
94. Italy	1.024	119.296	
95. Netherlands	1.461	64.585	
96. UK	0.087	113.552	
97. Austria	12.238	6.358	
98. Finland		9.468	
99. Norway	1.706	6.729	
100. Portugal		5.765	
101. Sweden		10.051	
102. Switzerland		6.009	
103. Greece		12.197	
104. Spain	1.982	43.892	
105. Yugoslavia	3.691	7.422	
CENTRALLY PLANNED EUROPE			
106. Albania	2.200		0.450
107. Bulgaria	0.144	10.554	
108. Czechoslovakia	0.149	14.176	
109. German Democratic Republic	0.075	16.434	
110. Hungary	1.997	6.817	
111. Poland	0.550	10.582	
112. Romania	14.486	4.538	
113. USSR	458.948	4.400	80.558
OCEANIA			
114. Australia	18.956	8.518	0.177
115. New Zealand	0.163	3.351	
116. Guam		1.650	

Source: World Energy Supplies — United Nations, New York 1976 (1950–74) Department of Economic and Social Affairs Statistical Office Statistical Papers: Series J. No. 19.

Table B
ENERGY PETROLEUM PRODUCTS
(Quantities in million metric tons)

Country	Imports	Exports
AFRICA		
1. Algeria	0.177	1.795
2. Libya	1.254	0.342
3. Morocco	0.275	0.010
4. Sudan	0.748	
5. Tunisia	0.357	0.114
6. Egypt	0.540	0.648
7. Congo	0.169	
8. Gabon	0.006	0.487
9. French Equatorial Africa		
10. Angola	0.319	0.080
11. Zaïre	0.254	0.205
12. Ethiopia	0.052	0.095
13. Ghana	0.015	0.223
14. Ivory Coast	0.026	0.149
15. Kenya	0.203	1.104
16. Liberia	0.003	0.001
17. Madagascar	0.021	0.195
18. Mozambique	0.188	0.023
19. Nigeria	0.309	0.109
20. Senegal	0.755	0.057
21. Sierra Leone	0.025	0.001
22. Southern Rhodesia	0.525	0.003
23. Tanzania	0.155	0.195
24. Zambia	0.119	
NORTH AMERICA		
25. Canada	3.623	9.615
26. USA		
OTHER AMERICA		
27. Argentina	0.650	0.047
28. Bolivia	0.012	0.284
29. Brazil	0.589	
30. Chile	0.090	0.149
31. Equador	0.019	
32. Paraguay	0.038	
33. Peru	0.031	0.225
34. Uruguay	0.161	
CENTRAL AMERICA		
35. Costa Rica	0.113	
36. El Salvador	0.012	0.006
37. Guatemala	0.055	0.006
38. Hondoras	0.007	0.220
39. Nicaragua	0.041	0.004
40. Antigua	0.125	0.220
41. Bahamas	2.027	9.119
42. Barbados	0.153	

26

Table B *continued*

Country	Imports	Exports
43. Columbia	0.030	1.505
44. Cuba	1.850	
45. Dominican Republic	0.480	0.080
46. Jamaica	0.442	0.057
47. Martinique		0.174
48. Mexico	2.121	0.250
49. Neth. Antilles	4.390	34.203
50. Panama	0.030	0.510
51. Puerto Rico	1.198	3.353
52. Trinidad	0.097	14.401
53. Virgin Islands	0.269	20.487
54. Venezuela	50.445	

MIDDLE EAST

Country	Imports	Exports
55. Israel	0.356	0.629
56. Bahrain	0.051	10.283
57. Cyprus	0.158	0.002
58. Iran		9.265
59. Iraq	0.030	0.125
60. Jordan	0.001	0.007
61. Kuwait		11.278
62. Lebanon	0.278	0.153
63. Oman	1.622	
64. Qatar	0.134	
65. Saudi Arabia		18.999
66. Democratic Yemen	0.072	2.080
67. Syria	0.956	
68. UAE	0.417	
69. Turkey	0.253	0.688

FAR EAST

Country	Imports	Exports
70. Japan	20.908	2.114
71. Okinawa		
72. Bangladesh	0.110	0.005
73. Brunei	0.011	0.002
74. Burma	0.015	
75. Cambodia	0.085	
76. Sri Lanka	0.030	0.015
77. India	3.685	0.160
78. Indonesia	0.300	6.602
79. Korea	0.238	0.736
80. W. Malaysia	1.217	0.147
81. Sarawak	0.119	0.962
82. Pakistan	0.630	0.387
83. Philippines	0.568	0.101
84. Singapore	4.102	13.985
85. Thailand	1.095	0.157
86. Malaysia/Singapore		

Table B *continued*

Country	Imports	Exports
CENTRALLY PLANNED ASIA		
87. China		0.220
88. Mongolia	0.357	
WESTERN EUROPE		
89. Belgium	6.479	8,627
90. Denmark	10.148	2.239
91. France	4.828	9.845
92. West Germany	28.883	6.010
93. Ireland	2.927	0.425
94. Italy	4.677	19.897
95. Netherlands	4.949	31.655
96. UK	12.248	11.213
97. Austria	2.020	0.072
98. Finland	4.170	0.283
99. Norway	3.240	2.076
100. Portugal	1.337	0.264
101. Sweden	18.817	1.172
102. Switzerland	7.369	0.192
103. Greece	0.967	1.250
104. Spain	2.567	3.600
105. Yugoslavia	1.109	0.248
CENTRALLY PLANNED EUROPE		
106. Albania	0.020	
107. Bulgaria	1.515	0.001
108. Czechoslovakia	0.292	0.192
109. German Democratic Republic	0.032	1.757
110. Hungary	1.001	0.336
111. Poland	2.060	1.140
112. Romania		6.197
113. USSR	0.767	31.744
OCEANIA		
114. Australia	3.574	2.430
115. New Zealand	1.205	
116. Guam	0.176	0.048

Source: World Energy Supplies – United Nations, New York, 1976 (1950–1974) Department of Economic and Social Affairs Statistical Office Statistical Papers Series J. No. 19.

NOTES

1. "Oil's first century" was celebrated in the United States in 1959, the centennial of the drilling of an oil well by Colonel E. L. Drake in Titusville, Pennsylvania: *Oil's First Century* (Papers given at the Centennial Seminar on the History of the Petroleum Industry — Harvard Business School, 1959). Cf. "There is a polite controversy as to whether the first oil well was drilled in Canada or the United States. The honour seems to belong to a well dug near Petrolia in the Province of Ontario in 1858", J. B. Ballem, *The Oil and Gas Lease in Canada*, p. 4. R. J. Forbes' paper in the Centennial Seminar traces the development of the petroleum industry in various parts of the world before 1859.

2. Edith Penrose, *The Large International Firm in Developing Countries*; (The International Petroleum Industry); Neil H. Jacoby, *Multinational Oil*; J. E. Hartshorn, *Oil Companies and Governments*.

3. Jacoby, op. cit., p. 26.

4. C. Tugendhat and A. Hamilton, *Oil — The Biggest Business*, p. 72.

5. Ibid., p. 71 et seq.

6. Jacoby, op. cit.

7. The San Remo Agreement, or more precisely the Berenger-Long Agreement, which was linked to the San Remo Peace Settlement of 1920, provided "for a division of Iraq's oil resources between British and French interests, in a 75:25 per cent ratio. The exclusion of American interests prompted the latter to demand and, with the aid of the United States Government, to obtain a share of Iraq's oil resources. A result of these negotiations was the inter-company Red Line Agreement of 1928, in which private and governmental undertakings were closely interwoven", G. Lenczowski, *Oil and State in the Middle East*, p. 167; see also Penrose, op. cit., pp. 94–95.

8. H. Cattan, *The Evolution of Oil Concessions in the Middle East and North Africa*, pp. 2–3.

9. "The seven international 'majors' ranked by their crude oil production in 1966 are: Standard Oil Company (New Jersey), Royal Dutch/Shell Group, British Petroleum Company, Gulf Oil Corporation, Texaco, Standard Oil of California and Mobil Oil Corporation (formerly Socony Mobil). Compagnie Françaises des Petroles is much smaller than any of these, as well as smaller than a number of other US companies, but it very early had a share in Middle East oil and is for this reason often referred to as an eighth 'international major'.", Penrose, op. cit., p. 89.

10. A. G. Bartlett III et al., *Pertamina, Indonesian National Oil*, pp. 47–48.

11. Tugendhat and Hamilton, op. cit., p. 112; the figure for the Soviet Union is from L. M. Fanning, *Foreign Oil and the Free World*, p. 351, and for the Dutch East Indies from Penrose, op. cit., p. 59.

12. Penrose, op. cit., p. 78.

13. Jacoby, op. cit., p. 186.

14. P. R. Odell, *Oil and World Power*, p. 14.

15. Penrose, op. cit.

16. C. P. Issawi and M. Yeganeh, *The Economics of Middle Eastern Oil*, p. 16.

17. Penrose, op. cit., p. 17.

18. Jacoby, op. cit., p. 246.

19. Issawi and Yeganeh, op. cit., p. 158.

20. M. A. Adelman, *The World Petroleum Market*, p. 80; *World Energy Supplies, 1950–1974*, United Nations, p. 159.

21. Penrose, op. cit., p. 73.

22. M. Tanzer, *The Political Economy of International Oil and the Under-developed Countries*, p. 130.

29

23. Adelman, op. cit., p. 100.
24. Jacoby, op. cit., p. 123 and p. 149.
25. Ibid., pp. 126–128.
26. Ibid., pp. 138–139.
27. "The concept of 'posted price' was first developed in the US oil industry. It simply means that any refiner or agent wanting crude oil posted a price at which he is willing to buy oil from a specified field . . . In the Middle East, the first to start posting prices were the parents of Aramco which announced their postings at the end of 1950. Other companies followed suit . . . their purpose was to provide a basis (acceptable to governments) as a basis for calculating taxable profits." Z. Mikdashi, A Financial Analysis of Middle East Oil Concessions, 1901-1965, p. 169.
28. M. S. Al-Otaiba, OPEC and the Petroleum Industry; Z. Mikdashi, The Community of Oil-Exporting Countries.
29. J. A. Yager and E. B. Steinberg, Energy and US Foreign Policy, pp. 257–258.
30. R. First, Libya – The Elusive Revolution, pp. 187–212.
31. Z. Mikdashi, The International Politics of Natural Resources, p. 68.
32. R. F. Mikesell et al., Foreign Investment in the Petroleum and Mineral Industries – Case Study: "Foreign Petroleum Companies and the State of Venezuela" by G. C. Edwards, p. 107.
33. G. W. Stocking, Middle East Oil, p. 148 and pp. 150–151.
34. Anglo Iranian Oil Co. case, Pleadings, Oral Arguments, Documents, International Court of Justice, 1952, p. 216.
35. Stocking, op. cit.
36. Extract from the Declassified Portion of the International Petroleum Cartel Report of the Federal Trade Commission, reproduced in Hearings before the Subcommittee on Multinational Corporations, Committee on Foreign Relations, 93 Congress, II sess. on "Multinational Petroleum Companies and Foreign Policy", Report, Part 8, pp. 530–531.
37. Stocking, op. cit., pp. 249–250.
38. Ibid., p. 190.
39. Ibid.
40. Ibid., pp. 130–199.
41. Mikdashi, op. cit., International Politics . . ., p. 52.
42. H. Madelin, Oil and Politics, p. 16: List of Latin American state-owned oil companies shows that Argentina's Yacimientos Petroliferos Fiscales (YPF) was founded in 1922, and that Uruguay, Peru and Bolivia had founded state-owned companies in the thirties.
43. Madelin, op. cit., pp. 31–54.
44. G. G. Richards, op. cit., in Mikesell et al., op. cit., p. 117.
45. National oil companies, or state-owned oil companies or entities have been established by quite a number of non-OPEC countries. Thus: India – Oil and Natural Gas Commission (1956); Pakistan – Oil and Gas Development Corporation (1962); Burma – Myanma Oil Corporation (1963); Brazil – Petrobras (1953); Chile – Enap (1950). Very recently, a number of such entities have been set up: Norway – Statoil (1972); Malaysia – Petronas (1974); Bangladesh – Petrobangla (1974); Britain – BNOC (1975); Petro-Canada (1975).
46. W. Friedmann and J. Beguin, Joint International Business Ventures in Developing Countries, p. 30; M. A. Mughraby, Permanent Sovereignty over Oil Resources, p. 71 et seq.
47. G. G. Edwards, op. cit., pp. 117–120.
48. A. G. Bartlett et al., op. cit., p. 191.
49. Ibid., p. 193.
50. International Legal Materials, 1965, Vol. 4, p. 809.
51. United Nations Yearbook, 1965, pp. 333–334.

52. OPEC Bulletin No. 8, August 1968, p. 4.
53. Jacoby, op. cit., p. 251.
54. Mikdashi, *The Community* . . ., pp. 142–145; Al-Otaiba, op. cit., pp. 141–147.
55. Odell, op. cit., p. 94.
56. Madelin, op. cit., pp. 222–223.
57. Al-Otaiba, op. cit., p. 168.
58. The metaphor of "an earthquake" has been used by a number of writers to describe the impact of the developments of October 1973, thus: Jacoby, op. cit., p. 257; Madelin, op. cit., p. 223.
59. Nationalisations: Algeria – 51 per cent interest take-over in French companies, appropriation of the whole gas line system, and nationalisation of the whole of the natural gas resources (1971); Libya – BP (1971), Bunker Hunt (1973); Iraq – IPC (1972), Mosul Petroleum, Exxon & Mobil (1973); Kuwait – (1975); Venezuela – (1975).
60. H. S. Zakariya, "Convergence and Divergence between the Exporting Countries: The OPEC Experience", Paper presented at the International Colloquium on Petroleum Economics at the University of Laval, Quebec, October 1975.
61. Tugendhat and Hamilton, op. cit., p. 344.
62. Jacoby, op. cit., p. 270.

31

CHAPTER II

Multinational Oil Companies and Petroleum Development

In the new global energy situation, for the great majority of countries, being net importers of petroleum, there is strong incentive for petroleum development. For many countries, not previously involved in this field, formulating law and policy means breaking new ground; for others existing law and policy call for reappraisal and re-formulation in the context of a totally new global environment. The basic objectives are rapid and thorough exploration of prospective areas, the effective development of, and maximum ultimate recovery from, reservoirs which are discovered, and maximisation of benefit to the national economy from such development.

Petroleum development requires substantial capital, sophisticated technical and managerial skills, and marketing outlets. Traditionally, such capital, skills and marketing outlets were provided (in areas outside the socialist world) by multinational oil companies: the legal framework within which this was done was that of a "concession". Traditional concessions were increasingly viewed by host governments as a colonial legacy which worked to their disadvantage. The resulting pressure for change led to the emergence of a number of different types of arrangements, ranging from "joint ventures" and "joint structures", to "service contracts" and "production-sharing agreements".

A government engaged in formulating or revising its petroleum development policy has to decide on the extent to which multinational oil companies are to be involved, and the type of arrangement under which private operators, and in particular multinationals, are to be invited to commit their capital and their capabilities to petroleum development. The government has also to decide about its own role — whether it would limit itself to defining law and policy and performing regulatory functions, or whether it would involve itself in petroleum operations, through a state-owned or state-controlled agency such as a national oil company.

A comparative study and evaluation of the experience of other countries in working different types of petroleum development arrangements can contribute towards the formulation of "a practical framework for decision-making" for governments engaged in making a choice from among a range of policy options. Thus, before deciding upon the role to be assigned to multinationals, it would be useful to assess the record of performance of multinationals in petroleum exploration and development in different settings and to assess the considerations which influence their corporate decisions to invest in projects for petroleum exploration and development. It would also be useful to identify issues over which conflicts have arisen between host governments and multinational oil companies, and to examine the mechanisms designed to eliminate or reduce the possibility of such conflicts. Finally, it would be instructive in this context to examine the choices available to a government in terms of alternative frameworks for petroleum development.

EXPLORATION – POLICY AND PERFORMANCE

Before formulating its policy on the role to be assigned to multinationals in its petroleum development programme, a government should be clear about how multinationals generally view proposals for exploration, and how different groups of them view specific proposals for exploration in a particular area.

This approach would enable the government to deal with different sets of arguments which are presented for and against multinationals. Those who urge a central role for multinationals in particular in developing countries argue that:

Oil exploration in new areas is an undertaking that requires costly effort, and promises uncertain rewards . . . there are literally thousands of determined efforts that result only in dry holes . . .
Technical competence reflects the training and experience that is built up only in the course of repeated operations in many areas and under all possible conditions . . .
On the financial side [one] would stress two vital contributions of private enterprise – in providing essential capital and subsuming foreign exchange costs . . . the flow of capital required to finance oil operations is very large indeed . . . in effect, world wide operations of the industry enable companies to draw on their earnings in established areas to finance exploration and development elsewhere.

In the same vein, a United Nations report on financing petroleum exploration has argued:

Although individual exploration programmes inevitably involve an element of risk, it might perhaps be possible, on the basis of a series

33

of operations spread over a long period in different areas, to achieve a rate of return that would offset the risks of single operations . . . This method . . . can only be put into practice by companies or institutions operating internationally . . . it is the practice of such companies to rely on their own resources to finance exploration . . .[2]

A critique of these arguments points out that while it is correct that multinationals have substantial resources in money and men, which *could* enable them to efficiently carry out oil exploration programmes, the critical question is: to what extent international oil companies would want to explore in the developing countries?[3] This leads to the question: what considerations influence the decision by multinationals to undertake exploration (in developing, as well as developed countries). Among those which are important are: the reserves currently held by them, the ratio of these reserves to current production, needs of the marketing outlets served by them, diversification of sources of supply (in particular acquiring sources in politically secure areas) and a preference for sources of supply located close to expanding markets. A consideration which influenced the majors holding substantial reserves to bid for new concessions was to pre-empt discoveries by potential competitors, which could lead to an erosion of market control exercised by the majors. Thus, they were concerned "in bidding for new concessions, less with finding new oil quickly than in making sure it [was] in 'strong hands" and there were cases where they were "not after a concession to produce oil: they were after a concession not to produce oil".[4] The companies in the fifties and sixties had substantial financial resources for investment, and there were strong fiscal incentives provided by some of the home countries, principally the United States, to encourage them to undertake exploration. This led in the judgment of some analysts to aggregate over-investment in oil exploration and development as "pressures of national security and competition tend to drive each country and each oil company to expand *its own reserves*, even though *reserves in the aggregate* appear to be ample".[5]

The following factors appear to have stimulated investment in exploration:

1. Most countries seek a reliable energy base by encouraging the development of indigenous resources and by giving diplomatic and financial support to the foreign operations of locally based petroleum companies.
2. The concession agreements of host governments press oil companies to search for oil, under penalty of the forced relinquishment or nationalization of their concessions.
3. The tax systems of the United States and many other countries

encourage oil investments by treating intangible drilling costs as business expenses, granting percentage depletion allowances, and permitting an American oil company to reduce its American taxes by offsetting against profits earned on domestic operations many of the expenses of carrying on exploratory activities abroad.

4. Companies with ample reserves in current producing areas nevertheless explore new areas in order to diversify their sources of crude oil for political, locational or quality advantages. They seek to protect themselves against the possibility that rivals may discover large reserves of low-cost crude oil. Few, if any, of the more than four hundred firms in the foreign oil industry possess *all* of the reserves they desire, of the *types* and in the *location* needed.

5. The "bonanza effect" operates in the oil industry . . . the lure of the "big strike" is powerful.[6]

But even during these periods of "aggregate over-investment" the pattern of exploration and development investments of different groups of multinationals reveals that the bulk of investments went to established producing areas, and a very small proportion to new areas in the developing countries. The geographical distribution of capital expenditures on exploration and development, from 1947–62, was as follows:

For the 1947–62 period, total capital expenditures in the non-communist world for oil exploration and development equalled $73.5 billion. Of this, 73% was devoted to oil exploration efforts in the United States and another 6% to those in Canada, or almost four-fifths in North America. Venezuela received 7.2% of the total oil exploration effort and the Middle-East 3.7%. In the oil importing underdeveloped world, Latin America received 4.2% of the effort, Africa 2.9%, and the Far East 1.8%, for a total of one-eleventh of the world effort.[7]

It is further evident that a high proportion of exploration in new areas, in particular in the developing countries, was undertaken by "newcomers" — "independents" and state-owned entities.[8] In 1948, the seven majors had exploratory interests of 1.72 million square miles, or 67 per cent of the total concession area then subject to exploration; in 1972, the share of the total concession area, then estimated to be 9.3 million square miles, held by the seven majors had fallen to 28 per cent.[9] The proven reserves held by the majors grew at 9 per cent per year, those of the others at nearly twice that rate. The growth achieved by the majors was largely from the development of known fields.

The decision to invest in exploration is one which a multinational or the basis of the profitability of a project and in the context of its own overall situation (reserves held, reserves–production ratio, marketing outlets), the investment priorities directed by that situation, and the investible resources available to it. Thus, "a company that is 'long' on marketing facilities such as Shell is prepared to invest heavily in finding new crude however long on crude others may be; one without sufficient market power to dispose of its supplies, such as the French Compagnie Française des Petroles is investing more heavily than others need to in gaining entry to new markets".[10] It had been argued for example that:

> established majors would have little direct incentive to explore for oil in India under any reasonable set of conditions. This simply reflects the fact that they all had large quantities of already discovered low cost crude oil available in the Middle East, which would be backed out by any oil discovered in India. (In relation to their own annual production of crude oil in the Eastern Hemisphere, crude oil reserves of Royal Dutch Shell would last 57 years, those of Jersey 79 years, Standard of California 82 years, and Texaco 82 years) . . . The only real positive incentive to an established major for exploring for oil would be some kind of "tie-in" deal particularly the right to build new refineries which would utilize the discovered crude oil and hence not back out its present Middle Eastern supplies. The only other significant motive for exploring would be a negative one; that is, if it was felt that unless the company went ahead with oil exploration, the government or competitors would do so and would ultimately find oil which would back out the established majors present Middle supplies . . . the only companies with any strong incentive to explore for crude without "tie-in" arrangements would be those which lacked cheap external crude but already had a refining or marketing position within India.[11]

Indeed, it is pointed out that Burmah Oil Company Limited, the only sizeable company which had made significant efforts to explore for oil in India in a joint venture with the government was one which fell into the category of a company which had a refining and marketing position in that country, but was short on crude.[12]

There is also evidence to suggest that calculations of "political risk" tend to influence multinationals to accord priority to projects in politically secure areas. Thus, it is argued that:

> Contrary to the popular view, the large international companies are usually "risk averters"; this is why an overwhelming proportion of their drilling effort is concentrated on areas with proved oil prospects (like the Middle East) or in or near large Western oil-consuming countries, which are politically stable from their viewpoint. Poor oil-consuming countries with uncertain prospects for oil do not figure highly on their list of priorities. Even when they acquire concessions in those countries and dig a few holes (as Stanvac did in India — taking more than ten years to dig ten wells), the primary motive is to keep potential rivals away, rather than to find oil. With the changed

36

conditions in the oil industry over the past year, it is not unlikely that this attitude of the international oil companies towards exploratory work in oil-consuming poor countries will somewhat change, in order to reduce their dependence on Middle East crude. But the "political risk" (e.g., nationalisation or the country concerned joining OPEC) of such investment would always be high from the point of view of international companies. This explains the emphasis they place on exploration in Alaska or the North Sea where the "political risk" is low.[13]

An assessment of the possible role of multinationals in the North Sea in the context of the early seventies, however, indicated that despite the consumer-governments' interest in the development of sources of supply outside OPEC control, the multinationals' interest was likely to be inhibited by such considerations as their continuing interest in the established oil-producing countries. Reviewing this subject, a European analyst had observed:

> . . . in this respect one must note an important restraint. This is the fact that most of the exploration decisions in the new areas, and the exploration and development itself, are undertaken by the very same companies which are currently responsible for oil production and developments in the traditional producing countries. They have, of course, established a relationship which determines the profits of the oil companies for the next few years and it is one, therefore which will not be lightly cast aside — particularly as the companies still hold out hope that the final expropriation of their remaining assets will come later rather than sooner. Thus, given some possibility of further mutually profitable co-operation between the parties, the companies have some incentive to choose to go slower on the exploration for, and the development of oil resources elsewhere, than is required to break the oligopoly. Even worse, having been successful in their exploration and development efforts, they may choose simply to sit on the bulk of the new reserves in order not to disturb the controlled supply situation, given that continuing to supply the consuming countries within the framework of the latter ensures higher profits for the companies than could be made out of the use of the alternative resources, the very exploitation of which could undermine the system out of which the higher profits would emerge.[14]

The role of the multinationals must be understood in the context of their being enterprises basically concerned with maximization of profit over time from their overall global operations. Some insights into how a multinational oil company views petroleum development projects can be gained from the testimony of some of their leading executives. A senior Vice-President of Exxon has thus described their approach to the making of investment decisions:

In view of the relatively long lead time required to develop oil fields and put other Exxon facilities into place for refining, shipping and marketing, we must plan our investments at least three to five years ahead. In addition, we must try to predict the utilization of the investments over the longer period of their useful life. To be able to do this planning for Exxon facilities, we usually make a forecast of what demand and supply conditions will be for the entire industry. This requires that we make many assumptions not only about economic factors related to the industry, but also broad economic and political developments. Forecasting for periods at least two to five years ahead is bound to be an exercise fraught with uncertainties . . . Our forecasting is worldwide. It starts with forecasts made by our affiliates in each country where we operate of both supply and demand for energy and petroleum. The individual country forecasts are built into a balanced, world-wide forecast of energy and petroleum supply and demand.[15]

The Chairman of Socal in a memorandum prepared in June 1974 explains the considerations which constrained development of reserve capacity outside the OPEC countries in the sixties (considerations which may not be altogether irrelevant even today in the appraisal of new projects outside the OPEC countries) thus:

Each oil company seeks to diversify its sources of production. But implicit in the concept of the need for a "strategic spare capacity capability" is the assumed fact that the industry could and should have found and developed adequate reserve capacity in countries outside of the OPEC orbit to hedge against the embargo of the Arab nations and the price action of OPEC as a whole . . . The OPEC nations represent governments which collectively control over 85 per cent of the producible reserves in the Free World outside the United States. They include countries stretching from South America and Africa to the Middle East and South East Asia. It would literally have been impossible to find funds in an era of large surplus capacity to explore in countries outside OPEC, find oil, drill wells, instal producing and gathering facilities, and then let the whole investment stand idle against the possibility that during some future political confrontation, this additional surplus capacity might be used. Further, no nation would permit a concession area to remain unproductive. *Moreover, any nation in which a significant discovery was made undoubtedly would have joined OPEC.*[16]

In the new environment, multinationals are engaged in working out their policies towards petroleum development. As enterprises committed to realise certain objectives, the principal among them being maximisation of long-run earnings,[17] they will be guided by their own global forecasts and scenarios of the future. They will be influenced by their home government's policies which may provide incentives for greater involvement in domestic exploration. This may be particularly relevant for US companies.[18] They will also be affected by the emerging trend among host countries to assert "permanent sovereignty over natural resources", involving not only claims to a larger share of the economic rent, but control over all aspects of petroleum operations. The conditions of vertical integration and oligopolistic control, in which the international petroleum industry conducted its business being no longer operative, the integrated multinationals are called upon to reappraise their policies. The view has been put forward that:

> We might read the history of the oil industry from 1970 onwards as an attempt by the biggest companies to react to the situation and restore their profit margins . . . Vertical integration upstream now rests on a *de facto* situation, while the owners of crude, the producer countries, are still uncertain what form to give their activity. In the near future, they may be able to put an end to the whole idea of vertical integration of the oil companies, by attempting to control other phases of the cycle . . . These changes obviously encounter strong resistance from the powers that be. The integrated multinationals still control the producer countries' crude, strong as they are in their down stream integration, their logistic capacity and (in the case of the American companies) the political support of their government. The shift from upstream to down stream of the big companies' strong point is the sign of an advanced diverticalization of the oil industry . . . It is easy to predict that from now onwards the oil firms will pursue quite different strategies from those of the "fifties and sixties".[19]

What are these strategies likely to be? This question must be one of special interest to governments, which are involved in deciding upon the role to be assigned to multinational oil companies in petroleum development in their respective countries. Some new orientations in the policy-planning of multinational oil companies are already evident.

In the words of one of Shell's directors:

> . . . the expectation that they [the established international oil companies] are infinitely flexible in their capacity and willingness to invest in the supply of energy regardless of the controls, obstacles, and difficulties that may be put in their way is fallacious. Already changes in approach to the new situation are discernible. These are changes in organisation to respond to new relationships upstream; changes from a "supply" activity within an integrated framework to a "trading" and technical service activity in which ownership of crude production has diminished; responses to the expectation that oil opportunities may be limited but that the world will need energy from other sources; even that energy may be so dominated by

governmental control that it is better to seek to use the human and other resources of a company in other ways. The view that diversification into non-energy business is in some way "illegitimate" reflects a remarkably naive and curious concept of economic processes.[20]

The "new relationships" sought by the multinationals are based on their control over technology and managerial skills, over logistics and transport, and over marketing outlets. Their bargaining strength derived from these factors is expected to secure their position in established oil-producing countries, so that though they may have lost ownership of the resources and control over supply, they will seek to negotiate arrangements under which the producing governments will assure supplies in return for the companies' contribution in the spheres of technical skills, transportation and marketing. Thus the multinationals can strengthen their control over downstream operations, where they still have a strong position; thus:

An examination of the various stages of the petroleum exploration, production, transport, refining and distribution processes suggest that there are a great many areas in which international oil companies have unshaken foundations. Such seems particularly to be the case at the downstream end of the business . . . In the marketing and distribution area, the internationally integrated companies have been steadily expanding the extent of their direct control of distribution channels . . . More recently, the side effects of supply reductions by OPEC countries strengthened integrated international company positions in consumer markets relative to unintegrated refining and distribution companies . . . In refining . . . the eight largest companies, let alone other integrated refiners, probably did about three-fifths of Europe's refining . . . In the area of logistic and transport . . . the fact that about 85% of the world's tanker fleet is owned, or on long-term charter to, the seven international majors alone . . . gives them a flexibility unmatched by individual national governments, or even groups of national governments. If the overcapacity situation now forecast in the world tanker market forces some independent operators to withdraw from the business of transporting oil during the coming year, current crude supply reductions could again have the paradoxical effect of strengthening the relative position of international companies.[21]

Multinationals and, in particular, the majors are also responding to the new environment by diversifying into other energy fields. Thus, for example, Exxon has committed resources to coal and nuclear energy development and research in other energy areas[22] and BP has acquired interests in Canadian coal concessions and is bidding for them in Australia.[23] There have been acquisitions of coal interests in the United States by Gulf, Continental Oil and Occidental Petroleum.[24] Socal has acquired a 20 per cent interest in Amax, a leading US producer of copper, lead and zinc, which holds the fourth largest reserves of coal in the United States. The oil industry is said to account for nearly 50 per cent of the uranium reserves and 40 per cent of the uranium milling capacity of the United States, as well as nearly 33 per cent of the United States coal reserves.[25] Diversification beyond the energy field finds multinational oil companies applying their know-how to areas such as industrial engineering, petro-chemicals and mining in entirely new areas. Mobil has acquired a property company and acquired substantial interests in Marcor, a company involved in retail stores and container manufacturing.[26]

In this background the assumption that because of the increase in oil prices and the attractiveness of the prospect of finding oil outside OPEC control, there would be a significant increase in investment by the multi-nationals in petroleum development outside the OPEC countries, may turn out to have been mistaken. It is evident that with the increasing cost of exploration and development, and financial constraints, multinationals would be considerably circumspect about making investment in exploration and development in new areas. Future scenarios as viewed by those companies indicate a much greater degree of instability in relations with host governments, and the view is expressed that "the risk of mineral exploitation in less-developed countries is much greater".[27]

The evidence of multinational company behaviour since 1973 broadly indicates that after an initial spurt of activity by multinationals directed at obtaining new exploration rights, there is a definite tailing off and a marked reduction in new concessions being acquired through 1975 and 1976. A survey of the reports of new concessions, published in the *Petroleum Economist* between 1974 and 1976, show that in the immediate wake of the price escalation of 1973, a spate of negotiations resulted in a considerable number of new concessions. In Asia, Africa and Latin America some forty-five countried granted new concessions. Seventeen of the countries granted "concessions" for the first time. Of the agreements signed by the seventeen countries, fifteen were signed in 1974 and 1975, and only two in 1976. While forty-five countries had granted new "concessions" in 1974 and 1975, only six are reported to have granted them in 1976. It is also of some significance that the great majority of these "concessions, were awarded to independents or state-owned oil companies — of the forty-five only thirteen concessions were to majors. Of the seventeen concessions granted by the "new entrant" countries, only seven were to majors; and in nearly all the cases the majors were one among a number of grantees, which were granted "concessions" in adjoining areas.

The response to the invitation extended by Petrobras to multinationals for grant of service contracts for exploration in Brazil, though made after overcoming domestic political opposition, is reported to have had a poorer than expected response, and of the forty companies which showed initial interest, only five groups remained until November 1976, when the first award was made.[28]

While company decisions relating to petroleum development are taken by multinationals within the framework of an overall strategy, based on what might be characterised as *macro* and global considerations, decisions in relation to specific projects are influenced by *micro* considerations of profitability, perceptions of the degree of *risk* (including political risk) and by the legal framework within which a multinational is invited to undertake exploration and development. The legal framework is relevant not only because this affects financial return, that is, profitability, but because other company interests, such as control over operations, freedom to take decisions relating to levels of production, to pricing and marketing, are affected by it. There are distinct interests which multinationals and host governments respectively seek to secure when they are involved together in petroleum development. The legal framework is intended to provide a basis which sufficiently safeguards the interest of each to provide a mutually acceptable basis for collaboration. Experience indicates that there have been legal frameworks which have worked to the greater advantage of multinationals, and where the interests of the host government have not been adequately safeguarded. It is also evident that alterations of the framework to secure greater advantages or safeguards for the host government have faced a degree of resistance from multinationals, and where their bargaining strength enabled them to do so they either prevailed upon the host government to back down, or they themselves withdrew. An early example of a situation where companies withdrew in response to a change in the legal framework is provided by Egypt in 1948, where: "Several major companies — among them Standard of Jersey, Socony-Vacuum and Royal Dutch Shell — had carried out geological and geophysical work over parts of the Western Desert in the immediate post-war years, but an unfavourable legal situation created by the passage of the 1948 Egyptian mining law, along with a corporate law requiring majority stock offering to Egyptians and Egyptian directors caused the companies to withdraw."[29] A more recent instance of negative company response is reported from Australia, where Tricentrol in its 1973 report stated:

> Steps are being taken to reduce our commitments, since it is unlikely that the return allowed in the case of a successful discovery would

be commensurate with the risk involved. The staff in Australia have been directed to the very fast developing areas of South-East Asia.[30]

But then even in regard to South-East Asia, a company representative reported:

> . . . for 1976 and 1977 the present trend of reduced exploration activity appears irreversible, in view of existing world conditions and *local government policies* . . .[31]

The new legislation in Malaysia, in particular the one which would give Petronas, the national oil company, the right to buy a new class of "management shares" carrying majority voting rights, met with concerted opposition from the companies. Thus Exxon refused to commit any further expenditure, and stopped further exploration, and Conoco withdrew its drilling rigs which were operating in Malaysian waters.[32]

In order to appreciate how legal frameworks operate to safeguard different interests, it is important to identify the interests which multinationals and host governments respectively seek to safeguard, and the major issues over which their interests conflict, for it is those issues which are central in the multinationals–host government negotiations, in which through a process of bargaining a basis for collaboration is sought to be achieved.

MULTINATIONALS AND HOST GOVERNMENTS: INTERESTS AND ISSUES

A government's principal objective in involving a multinational oil company in petroleum exploration and development within its territory is to secure an investment of risk capital and the technical and managerial skills of a multinational to carry out as thorough and as rapid an exploration of its prospective areas as is reasonably possible, and upon any discovery being made to secure the necessary investment and the necessary skills to develop the reservoirs discovered in a manner which will ensure maximum ultimate recovery and yield maximum benefits to the national economy. The multinational's principal objective of maximisation of its long-run earnings from its overall global operations determines its global exploration strategy. A multinational which decides to acquire rights to explore petroleum has no doubt an interest in exploration and in the making of a commercial discovery. But there could, depending on the position of the company, be basic divergences in the interest between the government and the company on a whole range of issues, on which

43

decisions would have to be taken at each of the different stages of operations that are normally embraced by petroleum development arrangements — exploration, development, production and marketing. Multinationals have, therefore, traditionally favoured the concessionary system under which ownership of any petroleum discovered was vested in them securing for them the powers of management and maximum freedom of action (decision-making) in respect of operations at each stage. The pace of exploration, the amount of exploration work to be carried out, the expenditure to be borne for exploration, the programme of development after discovery, the rate and levels of production, pricing, were all matters which could be exclusively determined by the company: the "ownership" which was vested in them reinforced their claim to exercise such powers. The government's interest was essentially limited to enjoying a financial return, be it in the form of royalty, rents and taxes, or a combination of these. In the pre-Second World War environment, governments of countries which possessed prospective areas tended to be politically weak and lacking in the technical and administrative capabilities required to identify their national interest, let alone supervise or control the operations of the companies. They were content to have exploration carried out, and for some financial returns to accrue to governments upon discovery.[33] It was in the post-Second World War environment that governments began to acquire a higher level of awareness which led them initially to demand higher financial returns. It was in the late fifties and sixties that the awareness began to extend towards a greater understanding of the structure and operations of the world petroleum market and the international petroleum industry. Governments began to perceive the divergence of interest that was inherent in different situations between the company and the government in the sphere of petroleum development. The principal areas where such divergence became evident are considered below.

Rate and extent of exploration

Governments have a clear interest in maximising the rate of exploration. They would like prospective areas over which exploration rights are granted to be thoroughly and rapidly explored by the most sophisticated available techniques. They would like the company to which such rights are granted to undertake a substantial work programme and make the necessary financial investment for the purpose. For a company acquiring exploration rights, the considerations affecting their decisions regarding the rate and extent of exploration are more complex, and depend upon

44

the position of the particular company in terms of exploration rights and sources of supply held by it in other areas, the needs of its marketing outlets, and the funds available to it. A company "short on crude" in relation to its marketing outlets is likely to share the government's objective of rapid exploration; a company which has other sources of supply and acquires rights primarily to pre-empt rivals and to ensure that new sources discovered by competitors should not erode its market control, is likely to allow the areas acquired by it to lie idle or to carry on little or no work work — perhaps the bare minimum to give an appearance that some work is being done. Reference has been made to such instances of "formal compliance" with minimum exploration obligations by the companies in Iraq;[34] an illustration is provided by the "negative interest" of companies, which are "long on crude" in making new discoveries by the testimony of a company executive on the initial discoveries in Oman:

> I had recommended to the executive committee that although we had the opportunity to go into Oman that we shouldn't do it because we were unable to provide adequate outlet for our Aramco concession . . .
> Just at this time, the producing department brought in their geologist who had just come back from Oman, and he stated, "I am sure there is a 10 billion oil field there", and I said, "Well, then, I am absolutely sure we don't want to go into it, and that settles it". I might put some money in if I was sure we weren't going to get some oil, but not if we are going to get oil because we are liable to lose the Aramco concession, our share of the Aramco concession anyway, if we were going to back up any further on it by going into new areas.[35]

Companies further have to decide on the extent of resources to be committed for exploration in a particular area in the overall context of areas spread over a number of different countries, where they hold exploration rights. These areas make competing claims on the exploration budget of the company.[36] While the government accords the highest priority to exploration in the areas falling within its jurisdiction, a company allocates resources among the different exploration areas held by it on the basis of a global plan, in which priorities are accorded by reference to a range of different considerations. Proximity to marketing outlets, physical conditions affecting exploration costs, and degree of political risk, are among the considerations which would lead to a particular area being preferred to another. Thus, a company supplying Japan is likely to accord higher priority to exploration in Indonesia than in North Africa, while one supplying the West European market is likely to accord priority to the latter. It is likely the companies regard areas in the North Sea or Alaska more politically secure than in some parts of the Third World — and may thus accord priority to areas held by them in such politically more secure areas, even though areas held in the Third World may be equally prospective, and where costs of exploration may be the same or even lower. A company, therefore, while acquiring exploration rights aims to retain maximum freedom of action so far as rate and extent of exploration is concerned. It would like to retain those rights for as long a period of time as is possible with the minimum of obligation upon it so far as incurring expenditure or carrying out exploration work is concerned. In this way, it would have greater freedom to determine its priorities among the areas in which it held exploration rights in different parts of the world. Governments, therefore, learning from the experience of the established producing countries, have in the new generation of petroleum development agreements been incorporating a number of different mechanisms to limit the freedom of companies in this regard. Thus mechanisms are designed to ensure that companies acquiring exploration rights are deterred from "sitting on" these areas and that they commit themselves diligently to carry out exploration, to incur the expenditures necessary for this purpose, and to relinquish areas progressively, until the entire area would be relinquished if the agreed level of exploration activity was not maintained, or if at the end of a stipulated period of time, no commercial discovery was made. These mechanisms, which will be considered in greater detail in a later chapter, include: allotment of smaller areas to a larger number of companies (as against allotment of large areas to one or a few companies), time-limits for exploration, area relinquishment/reduction requirements, minimum work obligation, minimum expenditure requirement, diligent performance requirement, approval of annual budget and work programme of company, submission of data and information, supervision of exploration work and participation in different aspects of exploration by personnel designated by the government.

Rate and extent of development

Governments have a clear interest in rapid development of any reservoirs which are discovered. A company has no doubt an interest also in such development, but here again a company has to take this decision within the global framework of its world-wide operations. Many of the considerations discussed above which lead it to accord priority to a particular area over another area affect its decision to invest in the development of reservoirs discovered in one area in preference to others discovered in a different area. Thus an "intelligence report dated February 1967" relating to Iraq introduced into the Senate records testifies that:

There is every evidence millions of barrels of oil will be found in new concessions. Some of these new oil reservoirs have been discovered previously but they were not exploited because of the distance to available transportation, the heavy expense of building new pipelines and the fact that IPC has had a surplus of oil in its fields that are already served by existing pipelines. The files . . . proved that IPC had drilled and found wildcat wells that would have produced 50,000 barrels of oil per day the firm plugged these wells and did not classify them at all because the availability of such information would have made the company's bargaining position with Iraq more troublesome . . .[37]

Even in Saudi Arabia, where highly profitable discoveries had been made, controversies arose in the fifties regarding the rate of development.[38] A restrictive attitude on the part of companies towards development of areas in the North Sea, up to the early seventies, has been explained in terms of a perfectly rational strategy giving weight to relevant factors in a global context, thus:

Oil from the North Sea was thus only interesting in economic terms in relation to imported oil readily available at refineries in Western Europe at under $3 per barrel. This minimized most companies' interests in too rapid exploration and exploitation of the North Sea's potential. Instead, at this time they only wanted a build-up of knowledge of the North Sea potential together with the opportunity to experiment with offshore drilling and field developments in deep water — in preparation for their expectation of an increasing need for such offshore oil later in the century . . . In 1971, moreover, the international companies had yet another reason for adopting a relatively go-slow attitude towards North Sea resources. This was a result of their success in achieving a modus vivendi with the Organization of Petroleum Exporting Countries (OPEC) . . . The modus vivendi achieved was highly favourable to the companies in the short-to-medium term if one could assume the ability of the companies to pass on price increases to oil consumers . . . the companies had achieved an agreement with OPEC which not only left them in control of the international oil system for another decade, but one which gave them a positive incentive to get as much as possible of the oil they had discovered in OPEC countries out of the ground and away to the markets of the world in the period up to 1982 . . . This of course meant that their interest in the North Sea both in respect of a rapid development of its potential oil supplies and in respect of its potential for natural gas in that such gas production would substitute for imported oil — was even further diminished and their commercial interests placed even more firmly in favour of the development of the potential after 1982, again implying a 1970s premium on securing knowledge and on gaining experience in off-shore procedures. European governments either lacked or did not seek access to alternative intelligence and thus accepted the very strongly commercially motivated presentations of the interested oil companies concerning the "limited" potential for North Sea oil over the next decade.[39]

The drastic alteration in the relations between the companies and OPEC governments following 1973, as well as the price rise, explain a change in the attitude of the companies in the North Sea, but the position described above illustrates the divergence in interest that may exist between companies and governments with regard to the development of potential sources of oil within their territories. Under arrangements in which freedom is left to the company to decide whether to develop the discoveries it has made, there could well be a situation where despite making discoveries a company could *rationally*, within the global context of its operations, decline to invest in the development of the reservoirs which it had discovered.[40] Further, where such arrangements did not provide for furnishing the government with information and geological data which had been yielded by exploration, the government might not even become aware that discoveries, which might be worth developing, had been made. Different mechanisms have been devised and incorporated in new petroleum development arrangements, which will be examined in greater detail in a later chapter, to reduce the risk of discoveries being underestimated or not being reported by companies with a view to postponing development or reducing the rate of development. These mechanisms seek to extend powers of supervision and control over operations, by requiring all information and geological and geophysical data, including interpretations, to be furnished to governments, by requiring the company to invest in development where discovery is made in "commercial quantities" (elaborate definitions of what is meant by "commercial quantities" are included), and provision is made for annual work programmes and budgets to be submitted for approval so that the government can monitor whether the company is making its best effort, consistently with "good oilfield practice"; and failure to develop a reservoir in which petroleum in commercial quantities is discovered, renders the company liable to surrender the area in which the discovery is made. Certain agreements contain provisions premised on the possible divergence of interest between the company and the government in a situation where a discovery may not be commercially attractive to the company, but may provide the government with a valuable source of supply for its domestic needs, and thus enable it to substitute imports; in such cases, it is provided that the company should relinquish the area in which the discovery is made in favour of the government, which can then proceed to develop it.

Rate of production and level of output

The rate of production, and the level of output, are issues over which the interests of the governments and the companies can clearly diverge. Some of the different situations of conflict have been identified, thus:

> First, the host country and the foreign company may have different views with respect to the maximization of returns from the production of resources. These arise from differences in outlook relating to long-run prices or from the application of different rates of discount to future revenues. The government may want to limit the rate of exploitation in order to conserve reserves for future production or it may expect that the long-term trend in prices will be upward, or at least not downward. These companies, in turn, may want to maintain their share of the world market by expanding output in relation to demand or by reducing prices to meet foreign competition. Alternatively, the companies may want to maintain prices at the cost of larger exports while the government may want to expand exports. This general type of situation, in which the foreign company and the host government entertain different views with respect to the optimum policy for maximizing total revenues over time, may be distinguished from the second situation. In this the interest of the foreign company and the host country diverge with respect to the desirability of maximizing total revenues (regardless of the time path). For example, the foreign company may want to keep prices low for sales to its affiliates in other countries since the higher the price (and total returns), the larger the amount of total revenues which must be shared with the host government. The foreign company may also want to limit output from a relatively high-cost source of supply in meeting its marketing commitments abroad in favour of lower cost alternative sources, or it may for security reasons desire to conserve its potential supplies in politically safe areas.[41]

One of the continuing sources of friction between the companies and the governments in the established producing countries has involved company decisions relating to the rate of production, growth of production and levels of output. In the sixties, many of the governments pressed for expansion of output, while the companies "programmed" production, orchestrating different levels drawn from their different sources, leading to a situation where most of the governments concerned felt aggrieved. As an Exxon document published in 1969 observed:

> No known method of allocating the available growth is likely to simultaneously satisfy each of the four major established concessions: i.e., Iraq, Iran, Kuwait and Saudi Arabia.[42]

Some of the other concerns of governments have related to controlling the rate of extraction so as to obtain maximisation of ultimate recovery

49

and conservation. Thus, for example, arose the issue of efficient utilisation of gas which is found in close association with oil in underground reservoirs. Such gas can either be flared or re-injected into the reservoir which, while involving additional expenditure, ensures greater ultimate recovery from the reservoirs. The experience in the Middle East has been described thus:

> For the most part Middle East concessionaires have flared the gas. The concessions have customarily left decisions on such matters to the discretion of the concessionaires. The flares have been a constant reminder to the host countries that the gas may not be used in their best interests and they have sought to change such provisions.[43]

In the context of the North Sea, governments have expressed a strong interest in being able to control production. Thus Norwegian petroleum policy as presented to the Norwegian Parliament makes it clear that:

> In view of its desire for a long-term exploitation of resources and on the basis of an overall evaluation in a social context, the Government has decided that Norway should adhere to a moderate pace of extraction of the petroleum resources . . . it will be necessary to exercise direct control over the rate of extraction, so that any large scale discoveries are not exploited faster than the popularly elected institutions consider desirable, based on an overall evaluation.[44]

In Britain, in testimony before a Parliamentary Select Committee, officials while conceding that "it is the practice in a number of countries, including the United States, to have a regulation of production", submitted that the position in Britain in 1974 (before the 1975 legislation was enacted) was that:

> The only power that the Government had under the existing licences arises from the obligation to conform to good oilfield practice, and I think it is fair to say that alone does not give the Government a very strong or positive power to alter the rate of exploitation materially.[45]

In support of the 1975 Petroleum and Submarine Pipelines Act in Britain, it was argued that the powers which it conferred on the government to control the rate of production and depletion were essential since,

> if we have unfettered extraction in a world which is increasingly short of energy, the pressures will build up so much that we are likely to see a 40-year potential, without a proper conservation policy, reduced to extraction lasting for only 10 years.[46]

Governments in the established producing countries have moved to assume powers of controlling production through "participation" in different forms in ownership and management, details of which will be examined in a later chapter, and in certain cases through nationalisation. The new types of petroleum development agreements seek to reserve powers whereby the government can regulate production in the interests of conservation, or require the company to conform to "good oilfield practice". These new petroleum development arrangement provisions aim to limit the freedom of action of the company to determine rates of production and levels of output. These mechanisms are examined in greater detail in Chapter IV.

Transfer of skills and technology, training

The basic strength of multinationals in their relations with governments arises from their possession of technical and managerial skills. A government if it has to safeguard its own interest through effective supervision and control of the company's operations and ultimately with a view to undertaking operations itself, must develop its own capabilities. This involves training of its own nationals and a transfer of technology and skills to be effected by multinationals. Training obligations incorporated in the old concessions did lead to a considerable development of trained manpower in those countries, but it is reported that:

> The Middle East governments have been jealous of their rights under these provisions relating to employment. Despite the training programmes that the concessionaires have inaugurated, they have agreed that the companies have deliberately disregarded the spirit of the agreements, if not their letter. The companies they profess to believe have been slow to train their nationals for positions of responsibility.[47]

Despite the growing strength of the national oil companies in the OPEC countries, there is still a considerable gap to be filled with regard to technical capabilities in different areas. A preliminary survey of OPEC capabilities indicates:

> . . . the share of national companies and institutions in [petroleum] services is very low, varying around 40% in drilling activities, dropping to 20% in geological and geophysical surveys; and further dropping to 15% in fixing and boring well casing pipes and to 10% in well treatment and testing. As regards electric logging, the national companies have no part whatsoever.[48]

It is argued that the structure of the world petroleum market itself discourages majors from effecting transfer to skills to state-owned national companies for:

As real nationalisation proceeds, the state companies typically become competitors of the established international firms in their developed country markets. For this reason, the more experienced is the international firm, the less happy it is to share its skills with state enterprises . . . As long as producing-state companies are viewed by the most important international companies as potential disruptors of market shares and profits in the industry, the established firms will have a natural reluctance to transfer technology and management skills to Arabs and OPEC nationals.[49]

Most of the recent petroleum development agreements and legislation have sought to devise provisions to impose specific obligations on companies to train and employ nationals and various targets are stipulated so as to enable governments to monitor compliance by the companies with their obligations in this regard. Thus "some provisions set employment targets by prescribing the percentage of the total staff or of skilled or un-skilled personnel that must be nationals . . . an additional restriction is imposed by laying down the percentage of total salaries and wages to be paid or earned by nationals".[50] Other mechanisms include stipulation of minimum training budgets, time-tables for replacement of expatriate personnel by nationals, specification of grades at which such replacement must take place, etc. The range and efficacy of such mechanisms will be discussed in Chapter IV.

Sharing of financial returns and pricing

Sharing of the total net revenue generated by petroleum exploitation has been a constant source of conflict between governments and companies. The problem arises because of the substantial "rent" that is generated, "rent" consisting of "the surplus generated from the difference between gross revenues from the sale of the products and the current expenditures for the productive factors".[51] The government as sovereign over the territory in which the resources occur, and as is the position in the great majority of countries, also the owner of the resources, expects increasing proportions of this rent since the company is perceived as making "excessive profits". The company on its part expects substantially more than the going rate of return on capital or the equilibrium price for the services rendered in the form of application of technical and managerial skills. The company sees itself as an investor of risk capital and regards itself as entitled to a share of the "rent". The respective perception of governments and companies has been described thus:

Companies usually justify the existence of exceptionally high returns on a particular investment in terms of the high risk involved in the extractive industries. They point to cases of millions of dollars invested in geological surveys and exploration from which the companies have received no returns, and argue that these unsuccessful investments must be balanced against high returns from the successful ones. The host government however tends to look only at the level of company earnings from those investments that have proved to be successful and argue that these earnings should constitute no more than a reasonable return on invested capital . . . [52]

The forms in which governments seek to realise their share of the rent vary in the different types of petroleum development arrangements. Thus they take the form of royalties, rents, taxes, share in the profits, or a share of the production, or a combination of these. The procedures for computation of the shares under the different types of arrangements have also been a source of friction and controversy. It has been noted that:

> The major sources of ambiguities and misunderstandings concerning financial allocation have generally involved prices being used in the determination of gross revenue, the calculation of depreciation, amortization, and depletion allowances, and the deduction to be allowed for payment to be affiliated enterprises for purchases of goods and services and for interest on debt.[53]

A "tonnage royalty" system was relatively simple to administer. The need for monitoring financial operations and the accounts of the companies arose, once governments were to appropriate a share of "profits". For "profits" could be reduced by manipulating accounts so as to inflate costs, and deflate earnings. Different types of accounting practices adopted by the companies operating in the Middle East gave rise to controversies on just this ground. The issues over which such controversies arose included treatment of "fixed costs", "dead rent", "drilling costs" and the companies' overhead expenses, incurred at its head office.[54] Controversies dealing with accounting practices which resulted in reduction of the companies' tax liability centred on deductions claimed by companies on the basis of "losses which had arisen from transactions in overseas oil at artificial prices" and losses incurred in connection with "extraneous activities" outside the North Sea, led to legislation in the United Kingdom to eliminate "loopholes in the existing rules governing taxation of profits of the oil companies".[55] While costs and losses can thus be inflated through the adoption of certain accounting practices, earnings can be deflated by sales to affiliates at artificially reduced "transfer prices". The Middle East solution to this was to require a "posted price" to be declared which would be treated as the basis for taxation. Both the recent Norwegian and British legislation have sought to deal with the problem of "transfer-pricing". Mechanisms have thus to be devised to monitor

53

the financial operations and accounts of the companies so as to ensure that the government's share is not unjustifiably reduced. We shall examine these mechanisms in detail in Chapter V.

Legal provisions have thus to be devised to deal with matters on which interests of the government (as sovereign and, in most cases, owner) and the companies (as the contributors of risk capital and technological and managerial skills) diverge. It has been said that "a government with new petroleum fields to explore and develop must . . . form an alliance with an oil company".[56] The terms under which such an "alliance" can be formed are the product of negotiations conducted within the legal framework established for this purpose. A number of countries have, in the context of their own special circumstances, undertaken petroleum development within a framework where capital, technology and skills have been secured not from multinationals but from other governments or state-owned companies. Governments are, thus, called upon to choose between different options and types of petroleum development arrangements. This choice must be made on the basis of a coherent strategy, which seeks to attain objectives and reflects priorities which are dictated by the particular situation of the country concerned. We turn therefore, in the next chapter, to a consideration of the elements which go into the making of a national strategy for petroleum development.

NOTES

1. W. J. Levy, "Basic Considerations for Oil Policies in Developing Countries", *Techniques of Petroleum Development: Proceedings of the United Nations Inter-regional Seminar on Techniques of Petroleum Developments, 1962,* New York: United Nations, 1964, p. 327.
2. United Nations, Department of Economic and Social Affairs, *Petroleum Exploration: Capital Requirements and Methods of Financing,* New York: United Nations, 1962, p. 15.
3. M. Tanzer, *The Political Economy of International Oil and the Under-developed Countries,* pp. 117-133 and p. 367 et seq.
4. J. E. Hartshorn, *Oil Companies and Governments,* p. 95.
5. N. H. Jacoby, *Multinational Oil,* p. 70.
6. Ibid., pp. 70-71.
7. Tanzer, op. cit., p. 131.
8. "In the international industry non-major companies are traditionally referred to as 'independents' ": see E. J. Penrose, *The Large International Firm in Developing Countries,* p. 133. They may be broadly divided into three groups: non-American independents, such as CFP, Burmah Oil, Petrofina (Belgium); American Independents and the state-owned companies, such as ENI, ERAP (France).
9. Jacoby, op. cit., p. 178.
10. Hartshorn, op. cit., p. 96 (setting out the position as it existed in the early sixties).

11. Tanzer, op. cit., p. 238.
12. Ibid., p. 238; R. Vedavalli, *Private Foreign Investment and Economic Development: A Case Study of Petroleum in India*, pp. 127–136.
13. B. Dasgupta, "The Changing Role of the Major International Oil Firms", in W. Widstrand, *Multinationals in Africa*, p. 75.
14. P. Odell, *Oil and World Power*, p. 231.
15. United States Senate, Committee on Foreign Relations, Sub-Committee on Multinational Corporations, "The Multinational Petroleum Companies and Foreign Policy, Hearings", *Report* (hereinafter referred to as *MNC Hearings Report*), 1974, Part 7, p. 334 (prepared statement by Mr George J. Piercy, Senior Vice-President of Exxon Corporation).
16. *MNC Hearings Report*, Part 7, p. 571 (prepared Statement of Mr Otto R. Miller, Chairman, Standard Oil Company of California).
17. Z. Mikdashi, *A Financial Analysis of Middle Eastern Oil Concessions: 1901–1965*, pp. 236–237; Mikesell, *Foreign Investment in the Petroleum and Mineral Industries*, pp. 35–36.
18. Yager and Steinberg, *Energy and US Foreign Policy*, pp. 390–391.
19. M. Colitti, "Vertical Integration, Multinational Oil Companies and Newcomers", Paper read at the Petroleum Economics Seminar, University of Oxford, March 1976, pp. 11–12.
20. G. Chandler, "The Role of the International Oil Companies", Paper read at the International Colloquium on Petroleum Economics on "La Gestion du Secteur Petrolier: De la Firme Multinationale Aux Rapports entre Etats", October 1975, pp. 8–9.
21. L. G. Franko, "Arab Countries and Western Oil Companies: Is Co-operation Possible?", Paper read at the Seminar on Administration of the Oil Resources of Arab Countries (sponsored by the Arab Institute for Social and Economic Planning, Kuwait), Tripoli, April 1974, pp. 1–5.
22. *MNC Hearings Report*, Part 9, p. 126.
23. *Financial Times*, 19 January 1977, p. 11.
24. C. Tugendhat, *Oil – The Biggest Business*, p. 371.
25. Ibid., pp. 372–373.
26. S. Fleming, "The oil companies prepare for the day when the oil runs out", *Financial Times*, 24 May 1976, p. 14.
27. M. A. Adelman, "Politics, Economics and World Oil", *American Economic Review*, Proceedings of the 86th Annual Meeting of the American Economic Association, May 1974, p. 63.
28. *Petroleum Economist*, 1976, Vol. 43, p. 4.
29. Fanning, *Foreign Oil and the Free World*, p. 245.
30. *Petroleum Economist*, 1975, Vol. 42, p. 6.
31. Ibid., 1976, Vol. 43, p. 106.
32. Ibid., p. 99.
33. Mikdashi, op. cit. (*Financial Analysis*, etc.), pp. 1–89.
34. *Vide* Chapter I, p. 00.
35. *MNC Hearings Report*, Part 9, p. 311.
36. This point was sharply focused in a reply given by Sir Robert Marshall, Permanent Under-Secretary, (Industry) of the United Kingdom to the Public Accounts Committee: "116. On the question of speed, is there not a slight variant in that the resources, particularly of the oil companies, were so limited that it was impossible to push them beyond a certain speed because they would not be able to raise the very great finance required for proper exploration of the sea bed? – (Reply): That is a world-wide problem and not one confined to the development of the United Kingdom Continental Shelf. There is great concern, I think, about the financing of the industry's operations, particularly again I think under

the pressures of OPEC because more and more attention is being paid to unconventional sources of oil. As for the United Kingdom Continental Shelf and British companies I think it is important to keep this in proportion. Total exploration effort in the free world on average during the 1960s was around £450 million a year. That is the total exploration effort by the oil industry in the western world, the western and eastern free hemisphere; and in the eastern hemisphere it amounts to just over £100 million, including the middle East, and in Europe and the Continental Shelf it was £30 million. So, as you say, there are great limitations to the resources available and a judgment has to be made by the oil companies as to where they want to place them." *First Report from the Committee on Public Accounts on North Sea Oil and Gas*, 1972–73, pp. 79–80. A 1974 estimate of petroleum exploration and development expenditures presents the following projection: "The international oil companies are faced with growing government intervention and with spiralling capital expenditure programmes. In the early 1960s the big oil companies provided roughly 90 per cent of their own finance. By last year this had dropped to below 70 per cent. Now, with growing official intervention over profit margins and massive capital expenditure programmes, it is expected to fall below 50 per cent. Even before the crude oil price rises, Chase Manhattan estimated that the oil industry would need to spend over one trillion dollars ($1000 billion) by 1985. It was assumed that $400 billion would come from outside. Today, this figure which works out at around $27 billions per annum must be considerably larger. The Middle East's recent oil cutbacks have forced the major oil companies to step up their search for politically less sensitive sources of supply. Citibank has put the total capital outlays on energy (one-quarter of which would be for oil) at $125 billions a year by 1980 . . . The rapid build-up of interest is being paralleled elsewhere in the world, specially in South East Asia . . ." "It's all a question of oil and money", *The Banker*, 1974, Vol. 124, pp. 278–279 (in which also special problems faced by the bankers in financing petroleum development are touched upon). For a more recent projection of the financial constraints on petroleum exploration and development, see A. Lambertini, "Energy Problems of the non-OPEC developing countries, 1974–80", *Finance & Development*, 1976, Vol. 13, pp. 24–28. William Hall, "The coming crisis in North Sea Finance", *The Banker*, 1975, Vol. 125, pp. 125–130.

37. *MNC Hearing Report*, Part 9, p. 228.
38. Donald A. Wells, "Aramco: The Evolution of an Oil Concession", in Mikesell, op. cit., p. 216 and p. 224.
39. P. Odell and I. Smart, *The Political Implications of North Sea Oil and Gas*, p. 00.
40. An example is provided by the substantial discoveries of natural gas made in Bangladesh by Shell in the sixties, when this area formed the eastern part of Pakistan; after the state of Bangladesh was formed Shell was pressed, as a condition of re-validation of its license, to agree to make adequate investment to develop the discovered fields. They ultimately decided to relinquish and withdraw, as they indicated that in view of other global commitments they were not in a position to make further investments in Bangladesh.
41. Mikesell, op. cit., p. 30.
42. *MNC Hearings Report*, Part 9, p. 215.
43. Stocking, *Middle East Oil*, p. 137.
44. Royal Norwegian Ministry of Finance, *Parliamentary Report No. 25* (1973–74), p. 17.
45. *First Report from the Select Committee on Nationalised Industries – Nationalised Industries and the Exploitation of North Sea Oil and Gas*, 1974–75, p. 16.
46. *Official Report, Parliamentary Debates*, House of Commons, 1974–75, Vol. 891, p. 523.

47. Stocking, op. cit., p. 135.
48. L. G. Franko, op. cit., p. 4.
49. Ibid.
50. M. M. Olisa, "Comparison of Legislation affecting Foreign Exploitation of Oil and Gas Reserves in Oil Producing Countries", *Alberta Law Review*, 1976, Vol. 14, p. 396.
51. Mikesell, op. cit., p. 34.
52. Ibid., p. 35.
53. David N. Smith and Louis Wells, Jr, *Negotiating Third World Mineral Agreements*, p. 59.
54. Stocking, op. cit., pp. 205–209.
55. *United Kingdom Offshore Oil and Gas Policy*. White Paper. Cmnd Paper 5696. 1974 paras. 8.15.
56. K. W. Dam, *Oil Resources*, p. 12.

CHAPTER III

Government and Petroleum Development Objectives, Options and Strategies

A government committed to petroleum development must adopt a strategy which would harness the maximum of resources needed for this purpose on terms which would ensure that if petroleum were discovered, the maximum of benefits resulting from such discovery would accrue to the national economy. To formulate a strategy, objectives must be identified, as also the resources required, and the alternative sources from which these might be secured.

The objectives may be broadly summarised as follows:

1. Rapid and thorough exploration.
2. In the event of a significant discovery being made, rapid and effective development.
3. Production at rates and using methods which will ensure maximum ultimate recovery.
4. Pricing and marketing to ensure that maximum benefit accrues to the national economy.
5. Transfer of technology and skills to nationals.
6. Maximisation of over-all benefit to the national economy from petroleum operations, e.g. support to domestic industries, employment of nationals, etc.

Among the resources required are risk capital, technological and managerial skills, and marketing outlets. It is likely that the bulk of these resources would be provided by multinational oil companies. In order, however, to obtain these on the best terms, governments must be aware of alternative sources which they may be able to draw upon, and of the practical steps that they can take to improve their bargaining position in negotiations with the oil companies.

THE BARGAINING FRAMEWORK: A DYNAMIC MODEL

Traditionally, multinational oil companies provided the resources necessary for petroleum development on the basis of a "concession" granted by the government. As described in Chapter I, "traditional concessions", in particular those granted before the Second World War, appear in retrospect to have contained terms which were specially advantageous to the companies. Under these concessions companies obtained extensive areas (in some cases covering the entire territory of a country as under the Kuwait concession of 1934, and the Qatar concession of 1935, or nearly half a million square miles as under the Saudi-Arabian concession of 1933), for long periods of time (ranging from 60 to 75 years). Companies secured ownership of the petroleum that was discovered, appropriated the bulk of the revenues (originally paying only a fixed per ton royalty) and exercised nearly total control over all phases of petroleum operations.[1] These concessions were the product of a global environment in which seven or eight vertically integrated major international oil companies, operating in an oligopolistic world petroleum market, dominated the international petroleum industry. In negotiating with governments, their bargaining power (often backed by that of their home governments) was as strong, as that of the host governments was weak. Governments had no option but to grant concessions to the majors, on *their* terms.

The pattern of government–oil company relations, under the concessions, has undergone drastic alterations in the last 30 years. These have been brought about by changes in the global environment, in particular in the conditions prevailing in the international petroleum industry and in the consequent shift in the bargaining positions of government and oil companies. These changes have also resulted in the development of new types of legal arrangements between governments and oil companies, under which governments have been able to obtain the resources of the companies for petroleum development on significantly better terms. Among these new arrangements which will be analysed in the next chapter, are: "joint ventures" and "joint structures", "production-sharing" contracts and "service contracts".

The changes of the last few decades in the terms of concessions have led to the view that "a concession must be seen not as a contract, but as a process in which rights and obligations of both parties shift over periods of time as specific factors change".[2] Bargaining positions can shift due to changes in the structure of the industry, as was the case in the early sixties, when the entry of a large number of so-called "independent"

59

and state-owned companies eroded the dominance of the majors, led to competition for acquisition of exploration rights, and thus enabled governments to obtain improved terms under new types of legal arrangements. Structural changes in the industry also explain the success of the nationalisations of the seventies, where the Iranian nationalisation of the fifties failed. Shifts in bargaining positions can also take place due to purely local developments. An illustration is provided by the case of Peru, where the growth of the domestic market for oil in the sixties reduced its dependence for marketing on the foreign companies, and thus influenced its decision to take over the principal foreign oil company, the International Petroleum Company in 1968. The change that followed soon after is described thus:

> Discoveries of additional oil in the 1970s, however, changed the situation. This oil had to be exported from Peru. As a result, new invitations were extended to foreign investors. The strength of foreign companies was restored as their marketing inputs were again needed. With some assurance that their marketing contribution would leave them a number of years of profitable operations, foreign firms were willing to enter Peru.[3]

Shifts in the relative bargaining positions of governments and oil companies result not only from such changes in the global or the local environment, but from their changing role over the life of a concession from one phase of operations to another.[4]

It is important to recognise that a petroleum development arrangement is one which involves distinct stages of operations. Thus, there is a pre-exploration or reconnaissance stage, an exploration stage, a development stage and then a production/exploitation stage. It is important to note that the respective interests of the parties in each stage need to be identified, and for it to be understood that the balance of bargaining power shifts during each stage between the government and the company.

The shifts in bargaining power in each stage of a mining development arrangement have been outlined thus:

> [In Phase I], the bargaining strengths of parties at [the pre-exploration stage] depend on the amount of geological and other information available . . . Where basic information has to be obtained by the explorer, the government's position is usually weak. Where basic geological information is available and shows favourable indication, the government's position may be strengthened by the appearance of a number of competing prospectors. In such circumstances exploration rights can be allocated by tender, or auction, securing an initial return to the government. Even at that initial stage, economic and political factors as well as geological will be important.

[In Phase II] initial exploration has yielded promising results and an intensive programme to prove a commercial resource is mounted. A major feasibility study may be called for which requires expertise of a kind that at present is scarce and concentrated in the hands of the multinational mining companies, but there are independent operators whose services can be commended if finance is available.

[In Phase III] the construction (development) phase of the project major investment is required for development of the resource. At this stage political factors tend to assume a priority role. In the case of a major project a substantial part of which needs to be financed by loan, only a limited level of political risk acceptable, this trade-off is limited because large-scale lenders require assurance that the interest on their loans and repayment of their capital are secure.

[In Phase IV] in the early production stage, the loans needed for development of the project are amortized and investors recoup part of their original investment.

[In Phase V] . . . the project has matured and is no longer mortgaged to lenders and it is at this stage that the scale of bargaining strength tilts strongly in favour of the government. Basically all that's left in the weaponry of the private investor is the threat of withholding further investment or withdrawing his technical expertise in project management or in marketing.[5]

No doubt the relative bargaining strength of a company is greatest at the earliest stage where there are only broad indications of the possibility of the existence of petroleum resources. For the government at this stage there is a clear interest in attracting the capital and technology of the oil companies to undertake systematic exploration, involving geological and geophysical surveys, followed by exploratory drilling. The discovery of petroleum in commercial quantities significantly alters the relative bargaining position of the parties, since:

the perceived level of risk associated with the enterprise declines precipitately. The returns to the foreign company no longer seem appropriate to the risk, and the governemnt feels justified in demanding downward adjustments in the investor's share of profits. Meanwhile, the disposition of the investor to accede to marginal pressures, if this will protect the investment, tends to grow.[6]

At the same time, the willingness and ability of the company "to disengage from the venture without first obtaining back the expected remuneration diminishes as his stake increases through the infusion of equipment, personnel and capital".[7]

The relative weakness of the government's bargaining position in the earliest stage leads companies to concentrate on obtaining an overall agreement securing for themselves the best possible terms in all phases of the operations. While committing themselves to certain minimum work and investment obligations in the exploration phase, companies secure from governments commitments on the right to exploit and financial returns to them for the entire period of the contract (which may extend to a period of 20 years or more). From a government's point of view, it would, theoretically, be most advantageous to negotiate the entire package of terms after the exploratory work had been completed and the magnitude of the discovery evaluated.[8] A company, however, would rarely be willing to invest in a substantial exploration programme, unless it first obtained from the government a commitment that in the event of significant discovery it would have the right to exploit the discovery and earn substantial profits. It is, therefore, important to keep in view that companies

> attempt to capitalize on their initially superior bargaining position to secure the best possible terms. Any unnecessary compromise with the host country during the pre-contract bargaining period leaves the company with a corresponding loss of room within which to maneouver in the "next round".[9]

This underlines the importance for a government to approach the initial pre-contract negotiations with adequate preparations and a carefully considered strategy. The cynical view which is sometimes expressed by company and government representatives alike is that one need not be too weighed down by these negotiations and their outcome, since regardless of what agreement they may sign, the terms are likely to be altered subsequently, when bargaining positions shift. While this view may find some justification from the recent history of revision, and even abrogation, of concessions, there are weighty reasons why a government would be ill-advised to underestimate the importance of the initial negotiations and their outcome. Unless the government has made adequate preparations, dealing with the kinds of matters discussed in this chapter, and has formulated an overall strategy it may find that it has substantially prejudiced its position. Not only may it fail to obtain the best possible terms, but it may foreclose a range of other options for itself, as would be the case for example if it did not reserve certain prospective areas for subsequent rounds of licensing. Further, the extent of the company's obligations during the exploration phase, and the size and character of the exploration programme to be undertaken by it, are the result of the initial negotiations. A comparison of petroleum development arrangements indicates that there is a wide range of variation in the provisions spelling out the company's obligations during the exploration period. These range from the weak formulations, found in earlier agreements, which only cast an obligation "to commence geological and/or geophysical exploration" and to "drill a test well" within stipulated periods,[10] to those which have been more diligently negotiated, aided by greater information obtained as a result of geological and/or geophysical surveys, and which stipulate in specific and precise terms both minimum expenditure obligations and minimum work programmes, and, in some cases, provide for a bank guarantee to be furnished by the company guaranteeing due performance of its obligations.[11]

The success of members of OPEC in revising, or abrogating concessions should not lead to the impression that a government may unilaterally alter agreements, without facing a host of adverse consequences, even though its claim to have a legal right to do so may have an arguable basis. That the exercise of such a right even by an OPEC member exacts a substantial price is evident from the recent experience of Indonesia, which upon altering certain terms in the prevailing production-sharing contracts in 1975, in particular changing the 65:35 production sharing ratio to 85:15, was faced with a situation in which companies held back from undertaking exploration work in new areas, thus creating pressure which led the government to effect modifications acceptable to the companies.[12] Reference has been made in the last chapter to similar instances of withdrawal and suspension of operations by companies in response to the stiffening of the terms governing petroleum exploration by Malaysia and Australia. In this background, for a non-OPEC developing country, it may not be realistic to assume that the making of commercial discovery would so significantly improve its bargaining power as to enable it to rewrite the original agreement; though this might be the case if a number of large discoveries were made, in which event the country concerned may even qualify for OPEC membership. Modest discoveries made in developing countries, most of which are aid-recipients, are not likely to so increase their bargaining power, as to enable them to rewrite agreements, without facing adverse consequences, including reduction in aid flows.

Companies continue to project the view that one of the principal considerations that keep them from undertaking a foreign exploration venture is the risk of unilateral government action directed at the terms of the agreement under which the venture is established.[13] Circumstances which could contribute towards the stability of contractual terms are, therefore, positively rated by them. One such circumstance would be a legal regime, or agreement, which is carefully worked out, and reflects an understanding of the dynamics of the petroleum industry and the economics of petroleum development. This would also result in improving the domestic environment for investment in petroleum development, for it is suggested that:

If the governments of under-developed countries had more equal access to international expertise from the initiation of a concession (through training programs set up for this purpose and/or through hiring consultants to aid their side of the negotiations), it would be less likely that an entire political tradition would be built upon frustration and suspicion. More open negotiations, clearer expectations about costs and benefits, surer understanding of occult business practices might result in more confident acceptance of the foreign investors' presence. In all likelihood, the cycle of negotiation and re-negotiation in accordance with shifting bargaining power would continue, but the premiums and discounts would become standard and more palatable . . .[14]

FORMULATION OF A COMPREHENSIVE GOVERNMENT STRATEGY

The formulation of an effective government strategy for petroleum development must be based on an assessment on the one hand of the geological prospects of its territory and continental shelf and on the other, of the resources, technological and financial, which can be mobilised for exploration and development. Since the financial and technological resources required are normally beyond those possessed by government, it is necessary to identify alternative sources from which such resources can be obtained. The principal source for providing these resources traditionally and to a very large extent even today, are the multinational oil companies. There are, however, today certain other sources — state-owned national oil companies, government entities and international and regional organisations — which a government can draw upon, if the multinationals are not forthcoming. To the extent to which such alternative sources exist, and are further developed, governments can by judiciously drawing upon them, reduce their dependence on multinational oil companies and in any event, improve their bargaining power to obtain better terms from them. Since the terms are to be determined by bargaining, a government in preparing for negotiations, should aim to enhance its bargaining power and capabilities. The practical steps that it can take for this purpose may be considered under the following heads: (I) Administrative Organisation (II) Corporate Strategies and Government Options (in the context of conditions currently prevalent in the international petroleum industry) (III) Assessment of Geological Prospects (IV) Disposition of Prospective Areas and Demarcation of Blocks (V) Alternative Procedures for Awarding Areas and Selecting Companies.

Administrative organisation

Most governments, in particular in developing countries, are handicapped by a lack of expertise and experience, and face the problem which is described in the following terms:

> A country that has no history of large natural-resource concessions begins in all likelihood, with a very inexact knowledge of its mineral or petroleum wealth, and has very little independent capacity to check the feasibility studies or exploration reports presented by foreign investors (or even by foreign consultants). There has been no occasion to build up a bureaucracy informed about standard business practices in a particular industry, or familiar with the terms of concessions in other countries. In many cases a host government may not be experienced enough in handling transnational corporate accounting (transfer prices or inter-subsidiary financing) or international tax provisions to make the negotiation process very meaningful. In much of the Third World before the Second World War — even in regions not formally colonized — initial investments in resource extraction were undertaken with only the most primitive attempts at bargaining. When the international oil companies first approached General Vicente Gomez of Venezuela or King Ibn Saud of Saudi Arabia, they were invited to draft their own petroleum legislation.[15]

A government must, therefore, muster together if need be from outside the ranks of the civil service and government departments, qualified specialists who have the expertise and capacity to deal with the technological and economic issues which are involved in petroleum development. The creation of national oil companies is in many countries a response to the felt need to have a national machinery capable of dealing effectively with the multinational oil companies. Thus, reviewing the development of Pertamina, the Indonesian national oil company, it is observed that given the objective of effectively dealing with the multinationals

> the domestic enterprise would not only have to thoroughly comprehend an enormously complex industry, but would also have to assemble the resources necessary for competing with the companies, which participate in the industry. The latter element is a prerequisite if the local enterprise is to gain an upper hand in dealing with those over whom it expects to govern. In short, Pertamina has realized that in order to deal effectively with international oil companies it is necessary to *become* an international oil company. Thus, Pertamina has created an infra-structure which increasingly resembles that of an international company.[16]

The establishment of Statoil in Norway, and that of the British National Oil Corporation, are in part influenced by the need to develop national capabilities to deal more effectively with the multinational oil companies.

Special attention needs to be accorded to the formation and mainten-ance of effective negotiating teams. On the basis of actual experience, it is observed that governments have been prejudiced in negotiations with companies through such practices as frequent changes in the composition of the negotiating team (even from one session to the next), and failure to designate a properly empowered chairman of an inter-ministerial team, which provide company negotiators with advantages, such as the following cases illustrate:

(a) Where members of the government team are changed frequently, the company negotiators having gained acceptance of certain terms by the first set of government negotiators, can seek to have certain others, unacceptable to the first, accepted by the next team, who may be unaware of the implicit trade-offs made in the previous meeting;

(b) Where company negotiators can negotiate directly with individual ministries (as distinct from a unified government team negotiating on the basis of an agreed brief), and obtain concessions from them directly, the government side loses the advantage of making sophisti-cated trade-offs that cross ministerial lines of authority;

(c) Unless the chairman of the government team has the authority to ensure that a unified government position is maintained, company negotiators can exploit differences among members of the government team, playing off one against another, and thus erode the strength and effectiveness of the government team.[17]

In order to maintain the effectiveness of a government negotiating team, it has been urged that certain guidelines may be followed so that:

(1) their membership, no matter how it is made up, does not vary from negotiating session to negotiating session;

(2) they have a clearly designated chairman with clearly designated powers;

(3) they have unambiguous authority from the government to conclude agreements, subject only to executive or legislative approval.[18]

Corporate strategies and government options (in the context of conditions currently prevalent in the international petroleum industry)

Government strategy has to be formulated in the context of the global environment outlined in Chapter I, and the patterns of multinational cor-porate behaviour reviewed in Chapter II. The expectation that the OPEC-led changes of 1973 would stimulate extensive exploration in new areas cannot be said to have been fulfilled. Although there were a number of agreements signed in 1973 and 1974, for exploration in new areas, a marked decline is evident in 1975 and 1976. The expected "exploration boom" in the developing countries has not happened.

It is evident from a global review of prospective areas made by the US Geological Survey that the bulk of these lie in developing countries. The estimates for the major regions of the world, including continental shelves to a depth of 600 feet, having sedimentary areas with petroleum potential are as follows:

	Square miles	Per cent
Developed economies, except USSR	7,916,000	30
Latin America	4,890,000	19
Africa (including Malagasy)	4,722,000	19
USSR	3,480,000	13
South and South-East Asia	2,993,000	11.5
Middle East	1,200,000	4.5
People's Republic of China	900,000	3
	26,101,000	100[19]

A review of global drilling activity in 1976, excluding the activity in the socialist countries, however, indicates the following distribution:

	Per cent
North America	36
North Sea	16
Latin America	12
South-East Asia	9
Africa	9
Middle East	9
Mediterranean	3
The rest	6
	100

It is evident that oil companies are tending to concentrate their exploratory efforts in developed countries. The reasons adduced by them in support of their policy range from their perception of greater political risk and risk of unilateral government action in developing countries, to the advantage of securing sources of supply in close proximity to the developed markets of North America and Western Europe.[20] Added to these are financial constraints, and escalating costs. As companies are increasingly drawing upon the banking system for financing, the international banking system can influence the direction of exploration investments: thus its readiness to support ventures in the North Sea is cited as one of the reasons explaining greater activity in that area.

In these circumstances, companies demand a higher DCF rate of return from developing countries. While companies express their readiness to undertake exploration projects on the basis of a 15 per cent DCF rate of return, the evidence suggests that they are seldom responsive to DCF rates of return of less than 20-25 per cent in developing countries.[21] Unless the gap between the minimum DCF rate of return acceptable to a company for undertaking a venture in a particular country and the maximum that a country is willing to concede can be bridged through negotiations, no agreement can emerge. Thus, it is suggested that in the sixties, ventures in India did not attract the companies in part due to the fact that,

> the government offered a rate of return on capital employed of about 12 per cent inclusive of taxes, which comes out at about 6 per cent net of taxes, whereas the companies were demanding a 21–30 per cent DCF rate of return . . .[22]

If a government is disposed to engage the resources of the oil companies for petroleum exploration, it must attempt to assess the DCF rates of return at which companies may be attracted. While this can only become clear in the course of negotiations, a government could fortify its bargaining position by computing a range of DCF rates of return, on the basis of realistic assumptions, keeping broadly in line with DCF rates of return which are being offered, or earned, in "similarly-situated" countries for "similar" acreage. This is a type of information that could profitably be exchanged by governments, and pooled through the efforts of regional or international agencies. The complexities involved in computing DCF rates of return, and in making inter-country comparisons, are examined in Chapter V; this, however, is a necessary exercise which must be undertaken by governments as part of preparations for negotiations.

Of the factors which may negatively influence company responses to possible ventures in developing countries, some are common to all non-OPEC developing countries, while some others may be relevant only in the case of particular countries. Among those that are common to all are: (a) political risk, or risk of unilateral government action; (b) financial constraints. Among those that are relevant to particular countries are: (a) terms which do not give companies the rate of return and/or degree of control over operations which they may seek; (b) geological prospects, which are not attractive enough for the companies, even though they promise discoveries which may meet the needs of the country concerned;

(c) large domestic requirement for petroleum, due to size of population or higher level of industrialisation, so that even if substantial discoveries are made, the prospects of there being an exportable surplus of petroleum are low.

An awareness of these factors can help to devise appropriate strategic responses by governments. Two types of responses should suggest themselves. The first would seek to design conditions so as to provide appropriate inducements to companies to alter their negative position. The other would seek to obtain resources for petroleum exploration and development, or at least for certain phases of petroleum operations, from sources other than the oil companies. Among measures that could be adopted under the first head are the following:

(a) *Re-designing package of terms*
The package of terms on the basis of which a company is invited to undertake a venture may be re-designed, so as to offer it a higher DCF rate of return. A provision which is particularly attractive to companies is one which enables them to recover their capital within the shortest period — thus, a quicker "pay out" period has been pointed out by the companies to be one of the particularly attractive features of the production-sharing form of contract.[23] This is not only preferred as it ensures a better financial return, but also reduces the risk of alteration of terms, following upon a "change of circumstances". The risk of such a change is less over a shorter period than a longer one. Other types of provisions which contribute towards maintaining "stability of contractual terms" (considered in greater detail in Chapter VI) are: systematic divestment or "fade out" provisions with the margins of compensation agreed to in advance in the initial negotiations (backed by amounts put aside from time to time in escrow in the hands of a third party);[24] periodic review of financial returns; re-negotiation of certain terms on the happening of certain specified contingencies; flexible taxation provisions, such as the British Petroleum Revenue tax, which provide for an equitable sharing of profits and ensure that the bulk of any "windfall" profits are appropriated by the government.[25] A more basic re-packaging would reduce the financial contribution, and thus the financial risk, borne by the companies. Even under existing arrangements, a higher debt–equity ratio would reduce the risk of direct exposure. A substantial reduction of risk would, however, be achieved if risk capital could be mobilised from other sources, so that companies would in effect become sellers of services. Under appropriately framed service contracts, companies could contribute their technological and managerial resources in consideration for

payment, which might include a share of the petroleum recovered. Suitable incentives could be incorporated for successful performance.

(b) *Third-party financing*
There is evidence that

> more and more capital for mineral and petroleum projects have been raised in advance from processors and consumers and/or from financial institutions through various kinds of factoring (that is, selling collection rights on long-term contracts to a financial intermediary at a discount).[26]

In the present global environment, various new possible sources of financing may be tapped. International institutions such as the World Bank and the International Finance Corporation, which had hitherto as a matter of policy not been inclined to finance petroleum projects have re-appraised their policy. The Bank has now begun to give serious consideration to projects for investment in petroleum development. It has advanced $150 million to India for this purpose, and associated itself as a party in a petroleum development agreement between Gulf Oil and the Government of Pakistan, under which the bank is to contribute towards the cost of development, in the event that a commercial discovery is made. Major consumer governments which have an interest in the discovery of new sources of supply could be persuaded to provide financial support to petroleum exploration and development projects. The Japan Petroleum Development Corporation represents a consumer government initiative to support its companies to undertake foreign exploration ventures. OPEC and OAPEC are potential sources for financing exploration and development ventures in developing countries, as they have emerged as substantial donors and are particularly sensitive to the need to relieve the strain placed on developing countries by the upward adjustment of oil prices.

(c) *Involvement of international financial institutions*
Companies have indicated that they value the involvement of international institutions, such as the World Bank or the International Finance Corporation, in foreign exploration ventures, not only for the financial contribution which they can make, but as "stabilisers" which could contribute to ease tensions in company–host government relations.[27] It is recognised that in a contractual relationship extending over a long period of time, situations develop which unless mutually adjusted can strain that relationship, and even precipitate a crisis. The role of an institutional partner is seen as that of a conciliator, which will actively involve itself in bringing about adjustments, when such situations arise. Such institutions are seen

as possessing the expertise, the influence and "the leverage" needed to deal with such situations more effectively, than a company could if it had to cope with them on its own.

(d) *Dispute settlement by international arbitration*

The fact that the existence of such dispute settlement provisions did not deter the revision, or abrogation of some major petroleum concessions, and in cases yielded awards which proved to be unenforceable, has led some to argue that "dispute settlement is more of a technical detail of minor weight".[27] On the other hand, it is argued that

> Dispute settlement by international arbitration . . . has the function of deterring a host country from making re-negotiation demands . . . such provisions, with the threat of the negative publicity of an ICSID (international) award and its effect on the credit-worthiness of a host country are still not so insubstantial bargaining factors when the TNE (trans-national enterprise) is faced with renegotiation demands of the country.[28]

It is pointed out that the great care devoted to dispute settlement provisions in two recent negotiations involving hard-mineral contracts entered into by multinationals, namely the Ok Tedi contract in Papua New Guinea and the Botswana-Bomangwato contract of 1972, indicate that companies attach considerable value to a provision for international arbitration as a method of dispute settlement.

While these are measures designed to induce companies to undertake ventures, developing countries must also consider options to deal with situations in which companies may have to be substituted, and resources for petroleum exploration and development may have to be sought from other sources. Some of the options that may be considered under this head are outlined below.

Use of specialised contractors for exploration

While the cost of petroleum development may be beyond the resources of a developing country, it may be possible for it to undertake certain phases of it. Thus, for example, the cost of exploration expressed as a ratio of the cost of development of a reservoir can be in the region of 1:50. It is estimated that whereas the cost of development of a medium-sized field may be upwards of $100 million the exploration costs would be in the range of $2–5 million. Cost estimates of exploration made in 1975 give a figure for on-shore exploration of about $350,000 per year for every 1000 square kilometres, and for off-shore exploration of about $600,000 per year for every 1000 square kilometres. A typical exploration campaign extends over a period of 5 to 6 years. Thus, the cost of a typical on-shore campaign in an African country extending over 5 years is expressed by the following figures:

Item	Cost (millions of dollars)	Percentage of total
Seismic prospecting	4.4	16.4
Drilling	15.6	58.4
Miscellaneous and overhead (including civil engineering, maintenance of bases, etc.)	6.8	25.2
	26.8	100[29]

A government can, therefore, undertake exploration by engaging specialised contractors, and meeting the cost from its own resources or from financial assistance which may be negotiated for this purpose. The services of specialised contractors would be available to governments both for seismic prospecting and for drilling. Indeed most of the oil companies use the services of such contractors for exploration. Some of the state-owned oil companies have also developed expertise which can be secured for this purpose. Thus, Statoil now has a subsidiary, Statex, which undertakes seismic prospecting on a contract basis, and has undertaken work in areas as far afield as Sri Lanka. Burma engaged the services of American drilling contractors, Reading & Bates, to drill exploratory wells off-shore, and met the cost from untied Japanese loans.[30] The expertise needed for supervision can be obtained under contract from some of the oil companies, specially the smaller independents, or from national oil companies. Such expertise can also be obtained under inter-governmental technical assistance agreements or from international agencies, such as UNDP.

Indeed this option may be the only one for a number of countries, whose problems are noted thus:

> The problem of exploration is thus raised for the remaining 40 countries which have neither the financial means nor the human capabilities to undertake prospection on their own and whose petroleum potential is not good enough to attract financing from outside . . . An initial step to improve this situation would be to synthesize for each country the existing data on the results of geological and geophysical campaigns and drilling, in order to make a preliminary evaluation of the petroleum potential. For such a project to be

fully beneficial, some supplementary operations, such as additional geological or seismic prospecting (for example, a stratigraphic wall) might be necessary . . . The amount of money involved in compiling such syntheses would not be enormous and if no candidates could be found to undertake and finance them, it is possible that they might be carried out with joint financing by groups of companies (multiclient studies). Governments themselves could also assume responsibility for part of the financing . . . Such projects could benefit from synthesis reports compiled on a regional level. By grouping together and comparing data concerning several countries, it might be possible to generate new guiding ideas.[31]

The object of undertaking such exploratory efforts by a government is to increase the geological and geophysical information available to it. If this information were to indicate the existence of structures which may contain hydrocarbons, the government would be in a stronger position to attract companies, or to raise finance for drilling and development.

Inter-governmental and regional co-operation and public sector development

A number of countries, including developing countries, have undertaken petroleum exploration and development through government entities like the Oil and Natural Gas Commission of India, or the Oil and Gas Development Corporation in Pakistan, or through state-owned national oil companies, like Petrobras in Brazil, Sonatrach in Algeria, the Myanma Oil Corporation in Burma, Petronas in Malaysia, or Petrobangla in Bangladesh. The technology and technical and managerial skill required by these entities have in many cases obtained under inter-governmental technical co-operation agreements, by India and some of the others from the Soviet Union and Romania, and by Algeria from France.[32] These agreements provide for technical assistance, as well as for supply of equipment; the foreign exchange cost is borne out of a loan provided for that purpose.

The national oil entities have also been able to obtain technological and financial resources for petroleum development from companies under different types of arrangements, which are analysed in the next chapter. There are, however, alternative sources which they can draw upon. Apart from obtaining some of the technological resources needed under inter-governmental technical co-operation agreements, they can draw upon international organisations — the United Nations, its regional commissions and other agencies — as well as on other national oil companies. Regional co-operation, such as that developed among the national oil companies of Latin America, provide a model for developing effective schemes for regional co-operation in the field of petroleum development, which could provide valuable resources for developing countries. Some of these alternative sources are considered below.

United Nations. The United Nations has a wide-ranging programme of co-operation with developing countries in the petroleum field. Its projects in the field normally include four major components, namely expert personnel, equipment, contract services and training. The assistance rendered by it range from seismic surveys conducted (under a UNDP Project) for the Government of Trinidad and Tobago, a similar project in Chile, establishment of the Petroleum Exploration Institute in India, and of the central laboratory complex of the national oil company in Bolivia, to training drilling personnel and supporting a drilling programme in Syria.[33] Many of these projects have been supported by contributions from UNDP. Advisory services and other types of assistance have been extended through the United Nations Centre for Natural Resources, Energy and Transport. In addition, different programmes of regional co-operation in the field of petroleum development have been promoted by the regional commissions of the United Nations, considered below under the heading of regional co-operation.

National oil companies. Technology and expertise in the field of petroleum exploration and development has been developed by a number of national oil entities of developed countries, such as ENI of Italy, ERAP of France, and Statoil of Norway. ENI and ERAP have provided technology and expertise to a number of developing countries under joint venture agreements and service contracts, some of these arrangements being regarded as significant innovations when they were made.[34] Some of the government oil entities of developing countries have also developed expertise which they are in a position to provide for petroleum exploration under different types of arrangements. Thus, ONGC of India is committed to providing such expertise under a joint structure agreement in Iran, and under a service contract in Iraq, while Braspetro of Brazil has undertaken to do so under a joint venture agreement in Egypt, and under a service contract in Iraq.[35] NIOC, the national oil company of Iran, has undertaken an exploration venture in Indonesia, while YPF, the national oil company of Argentina, is undertaking exploration in Uruguay, under a joint venture agreement with ANCAP, the national oil company of Uruguay.

Regional organisations of national oil companies. In Latin America, the regional association of national oil companies, ARPEL (Asistencia Reciproca Petrolera Estatal Latinoamericana) has provided a framework for co-operation among them. Within this framework, a number of technical assistance agreements have been concluded under which a

74

national oil company has been able to draw upon the technological resources and expertise built up by another. Technical assistance has thus been rendered by YPF of Argentina to CEPE of Ecuador under agreements signed in 1972 and 1974, by YPF to ANCAP of Uruguay (1974), and by Petrobras of Brazil to ANCAP (1974).

OAPEC (Organisation of Arab Petroleum Exporting Countries). OAPEC has established, in 1975, a Petroleum Services Company, with its head-quarters in Tripoli, which will establish subsidiaries to provide specialised services including seismic surveys, analysis of geological and geophysical data, and well drilling.[36] While these services are likely to be utilised primarily by OAPEC's members, the capabilities and resources that will be developed by the Services Company can in time be drawn upon by developing countries under suitably negotiated arrangements. Indeed both OPEC and OAPEC might also be persuaded to provide financial support for petroleum exploration in developing countries.

Regional co-operation. Regional co-operation in the petroleum field has been promoted, to a limited degree, by the regional commissions of the United Nations — ESCAP in South and South-East Asia and the Pacific, ECLA in Latin America, ECWA in West Asia and ECA in Africa. ESCAP has undertaken a stratigraphic co-relation between the sedimentary basins of the region. It is intended to prepare a map of the basins. Two co-ordinating committees, the Co-ordinating Committee for Joint Prospecting for Mineral Resources in South-Pacific Off-Share Areas (CCOP/SOPAC) have been established, the former in 1966, the latter in 1972. Surveys have been held at their initiative. Overall costs of surveys have been reduced, and broad area studies have been possible within this framework. ECWA has undertaken geological studies and assessment of technical manpower requirements in the region. ECA in a regional confer-ence in 1975 adopted recommendations urging regional co-operation in a wide range of matters. Thus areas identified by them in fact provide a useful catalogue of areas in which regional co-operation could be promoted in other regions. These include: (a) co-operation among members in the search for new oil/gas deposits; (b) periodic meetings of experts to exchange information on the petroleum industry; (c) preparation of feasibility studies for sub-regional projects; (d) co-ordinated plan for the establishment of refineries in the region; (e) survey of all existing training and research facilities and manpower requirements in the field of petroleum and the establishment of new centres and institutions; (f) strengthening of existing training centres and institutions; (g) establish-ment of a regional petroleum institute; (h) exchange of petroleum information and visits of experts; (i) formation of joint field prospecting parties to undertake inventories of oil resources; (j) joint ventures for exploration of oil fields common to more than one country and exchange of exploration data; (k) co-operation in the exploitation of common fields astride national boundaries by drawing up common plans for exploitation, exchanging exploitation data and establishing common production policies; (l) establishment of a regional documentation centre, for the collection, analysis and dissemination, on a continuing basis, of up-to-date information on all aspects of the petroleum industry; (m) establishment of a regional energy unit to provide advisory services.[37]

Institutional Entrepreneurship

International organisations and financial institutions can play a significant entrepreneurial role in promoting petroleum development in developing countries. They possess the resources needed to act as a "catalyst" to bring together from different sources, both in the public and private sector, technological and financial resources needed to support an enlarged petroleum exploration programme in developing countries. They can devise new types of arrangements under which such resources can be brought together, and in some cases actively participate in the programme or venture. As has been noted above, the participation of an institution like the World Bank or the International Finance Corporation, can itself induce a more positive response from companies. Initiatives by international organisations could similarly induce state-owned oil entities and public sector organisations to play a more active role in foreign exploitation, supported by institutional financing. It may be beyond the resources of a developing country to locate new sources which it can draw upon for petroleum development. Nor will it be in a position to offer the type of inducement or assurance of security as could be provided by international institutions, on the basis of which new types of petroleum development agreements could be designed. In order to undertake this role, an organisation needs to initiate a survey of different sources which could be tapped for securing technological and financial resources for petroleum development in developing countries, and hold exploratory discussions with the public and private sector agencies concerned. The aim of these discussions would be to develop new types of arrangements under which petroleum exploration could be undertaken in those countries, in which no exploration, or not enough exploration, was being done under the existing types of petroleum development arrangements. While the view projected by the multinationals that there is a scarcity of technological capacity for petroleum exploration may have some validity, it is equally true that there are many entities, including specialised contractors, small independent companies and national oil companies, that have some surplus capacity which could be utilised for petroleum exploration in new areas, provided imaginative entrepreneurship, such as might be expected from international institutions, could harness it within the framework of new types of arrangements.

Assessment of geological prospects

Governments can significantly improve their bargaining position by acquiring geological and geophysical information about their prospective areas.

The phases of exploration through which such information can be gathered have been described thus:

> Exploration proceeds from reconnaisance, to regional, to detailed phases. In the regional appraisal, some of the main factors taken into consideration are: area of sedimentary basin or geosyncline, maximum thickness of the sedimentary column, type of sediments, existence of structural and stratigraphic traps, and identification of favourable migration and entrapment conditions for petroleum. The detailed exploration aims to rather accurately locate wildcat wells where they are most likely to discover an oil or gas deposit.[38]

The processes and procedures which have to be adopted in each phase of exploration and their relative cost may be summarised as follows:

> The geological survey may require the use of geophysical prospecting methods, such as gravimetry and magnetometry, associated with mapping techniques that use airphotos. However, since all of these methods use a relatively loose grid, the expenditure remains relatively low, especially if the fact that these operations are not specific to petroleum but normally form part of the general inventory of natural resources in a country are taken into consideration . . . If the geological survey shows the possible existence of hydro-carbons, two further exploration steps are taken. First, geophysical — mainly seismic — prospecting campaigns are conducted, in an attempt to spot within the basin the structures that, in depth, may have captured and preserved hydrocarbons. Secondly exploration drilling is done, with the intention of verifying the presence of oil or gas . . . Roughly speaking, the cost of these three phases of exploration breaks down as follows: approximately 5 to 10 per cent of the total is spent on the geological survey, 15 to 30 per cent on the geophysical prospecting, and 60 to 75 per cent on the wildcat drilling. The different specialised techniques (photogeology, field geology, core drilling, gravimetry, aerial magnetism, seismic prospecting, drilling, etc.) are not applied successively; there are usually several campaigns for each one, because the data obtained in a previous study may lead to the study of a new zone or, the use of other methods for obtaining a better understanding of the soil structure.[39]

77

It should be understood that evaluation of geophysical data involves fine judgment, and has been characterised by those in the field as being "more of an art, than a science".[40] As exploration proceeds, earlier assessments need to be modified in the light of new information, and further lines of study and exploration have to be decided upon. The complexity, and the subjectivity, involved is evident from the fact that there are many instances when an operator has relinquished an area, only to be followed by another, who succeeds using a different concept or technique. Thus, Occidental, using the latest seismic and digital recording technique, discovered an estimated 10 billion barrel field in Libya in 1966, in an area which had been explored earlier and surrendered by a major international company.

An awareness of the nature of the exploratory process is important in order for a government to adopt a number of measures to safeguard its interest. It is urged that governments should not only require that all exploration data should be furnished to it, but also evaluations made by the companies. Under the earlier "production-sharing contracts" in Indonesia, the companies submitted to Pertamina the rawest form of data, and contended that under the terms of the contract they were not under an obligation to submit any "derivative report" which was how the technological evaluation report was characterised.[41] Later production-sharing contracts specifically require evaluation reports to be submitted. The government should then have an independent evaluation made; in cases where the government does not have this capability, it should utilise the service of independent consultants. The importance of this has been emphasised by those who have been involved with administering petroleum contracts.[42] Not only may there be a difference in the evaluation of the data between a company expert and an independent or government expert, but differing views may be held among them about the new lines of exploration that should be carried out, as well as the intensity of the exploration programme. While it may generally be true that

> during the exploration stage there is less likelihood that Contractors will pursue courses of action inconsistent with the interest of the host country,[43]

there can be a situation where the interests of the company and the government may diverge. Thus, where the geophysical data shows indications of a small reservoir or one containing only gas, a company may not be willing to proceed with drilling, while it might be in the interest of the government to insist upon it. Since the company's field operations depend

78

upon financial allocations from headquarters, budgetary constraints may lead a company to slow down or have a less extensive exploration programme, and in particular a smaller drilling programme, than geological indications merit. Relying on appropriately formulated provisions, a government in possession of the relevant data and an independent evaluation can more effectively exercise its power to "approve" work programmes, and their execution. Effective supervision can ultimately bring about results not only favourable to the government, but the company, for as the Norwegian Petroleum Directorate is prompt to point out, the Ekofisk field would not have been discovered when it was had it not been for its insistence that the company proceed with exploratory drilling up to the stipulated depth, when the company, guided presumably by considerations of economy, had served notice that they intended to stop drilling and to relinquish the area.[44]

The possession of geological and geophysical data by a government, before it sets about to negotiate with companies, is advantageous in a number of ways. It enables the government to devise an overall strategy as to how best to dispose of its prospective areas such as by demarcation of blocks. The issues involved in this exercise are considered under a separate heading below. The information relating to a particular block may be *sold* to the company to which the block is alloted. Indeed some general information can be sold to all intending applicants. With such information in its possession, a government can negotiate a more effective minimum work programme for exploration, and judge the adequacy of work programmes and minimum expenditure commitments offered by companies.

The present practice among countries broadly reveals three different approaches: there are governments which have acquired basic geological and geophysical information before negotiating with the companies; there are others which have issued non-exclusive exploration permits to a number of companies which have been free to conduct geological and geophysical surveys, with the undertaking that the data gathered will be submitted to the government; the third group consists of countries which have granted exploration and exploitation rights to companies without first collecting geological or geophysical information about the areas which they have alloted.

The first course appears to be the most prudent. The cost of undertaking such geological and geophysical surveys is within the means of most governments, and there are cases where an agency such as the UNDP has provided financial support for the conduct of a seismic survey. The technology needed for this purpose can be obtained on a contract basis from specialised contractors. The non-exclusive licence issued to companies is a method which has the merit that the work is done by companies at their cost, and is a method which a leading company consultant commends in the following terms:

> the wise law gives everyone a liberal privilege of reconnaisance of going and looking so long as he does not interfere with surface properties . . . To achieve this the right to reconnoitre can be open to all without any formal permit or it may be evidenced by a permit available on demand . . . a company should not be required to turn over the results of the reconnaisance investigation to the government.[45]

This method, however, is obviously more advantageous for the companies, since it ensures that they have as much if not more information than the govnerment before they are called upon to make any commitments. The consultant's recommendation that the company should not be required to turn over the results of its investigation is one which no prudent government could be expected to accept.[46]

Disposition of prospective areas and demarcation of blocks

Geological and geophysical data relating to prospective areas provides a basis on which a government can draw up an overall strategy for exploration. Experience indicates that it has been considered prudent not to allocate the entire national territory for exploration to a single company or group of companies, as was done under some of the traditional concessions. The practice in countries with sophisticated systems of petroleum administration is to allocate certain portions of the prospective areas for exploration by companies, while retaining the rest either for allocation in subsequent rounds or for direct development by the government or by way of national reserves. The argument against allocating the bulk of the prospective areas in a single round is that this forecloses the possibility of a government obtaining better terms from allocations in subsequent rounds, after some discoveries had been made, thereby enhancing the geological rating of the adjoining areas, which remained unallocated. An argument against granting the entire or the bulk of the territory to a single company or a group of companies is that experience shows that the allocation of a number of small areas to different companies generates an element of competition, and yields larger and more intensive exploration programmes over the entire area allocated.

Some of the arrangements which have been adopted for a rational allocation of prospective areas merit consideration. Iran's Petroleum Act of 1957 lays down that one-third of the total exploitable territory was to be conserved at all times as national reserves; that the entire territory

80

(excluding the portion that was covered by an existing agreement with a consortium of companies) be divided into districts each containing not more than 80,000 square kilometres (Article 5); only restricted areas would be granted to individual operators "in no case larger than 16,000 square kilometres, with a maximum of five districts to be operated by the same company".[47] The British Petroleum (Production) Regulations, 1976, prescribe that applications for production licences may be made for "one or more blocks", being areas demarcated with distinct reference numbers in a map lodged at the Department of Energy, and notified in the Gazette as blocks for which applications are invited (Regulation 7). It is laid down that an application cannot be made for part of a block. The circumstances in which the "block system" was evolved when the first round of licensing took place in the British sector of the North Sea 1964 has been described thus:

> An important issue in the first round was the size of the areas to be licensed. The decision of the Ministry . . . was to divide the North Sea into a series of small blocks. The blocks, rectangular except where the proximity of a coastline (or in the second round of licensing a boundary line) required an irregular shape, were approximately 250 square kilometres in area. They measured about 8 to 9 miles (east–west) by 11 miles (north–south). The size of the areas was a question of some significance. Most oil companies sought somewhat larger blocks than the Ministry chose . . .[48]

The demarcation into blocks enables the pursuit of a variety of objectives. Since blocks would vary in quality, the government could allot some of the better blocks to national companies, or reserve them for allotment later on better terms. It could consider tagging highly prospective blocks with some which were less prospective, thus ensuring that less prospective blocks were not neglected. Where there were a number of good blocks clustered together, government could allot alternate blocks, so that in the event of a discovery those that had been retained could be developed directly by the government, or be alloted at a premium. It has been suggested that in the first round of North Sea licensing in Britain "more than 30 per cent of the most sought after blocks went to British applicants" and that BP obtained "licences for perhaps the largest number of blocks in what geologists had always considered the most prospective areas . . ."[49] A further advantage of this system is that by inviting companies to submit work programmes, block-wise, the government is able to assess the relative assessment of the blocks made by the companies, since more prospective blocks are likely to attract more active drilling programmes. The government is also aided in making an assessment by inviting applicants "to submit a detailed statement of their relative preference as among different blocks and different combinations of blocks".[50] Norway has adopted a block system for purposes of licensing areas in the North Sea, though its blocks are approximately twice as large as British blocks. It is evident from Norwegian practice that the block system has enabled the government to adopt allocation policies designed to secure certain strategic objectives. A Report to the Norwegian Parliament for 1973-74 identifies some of these objectives:

> The Ministry of Industry proposes to make the first allocations of a limited number of blocks in the first half of 1974, as this will make drilling possible in some of the blocks in the same year. In particular it is considered necessary that some of the borderline blocks shall be drilled at the earliest opportunity, since problems of quite a special nature exist there . . . Several known structures extend across the borderline and drilling must be expected to start there in the near future. In such cases, a find on the British side could have the result that Norwegian petroleum reserves were drained from the British side; or that production on the British side caused pressure to fall in the Norwegian part of the reservoir, consequently reducing the recoverable reserves on the Norwegian side, or making production more expensive. It is, therefore, necessary to find out as soon as possible how many deposits there are that extend across the borderline, and determine how they are to be exploited . . . Priority will moreover be given to awarding blocks that can provide knowledge of conditions farther north and in deep water . . . (Statoil has selected certain blocks, which would be alloted to it) as parts of the efforts to establish and develop the company.[51]

A Report to the Norwegian Parliament for 1975-76 identifies a number of objectives to be kept in view when awarding areas for petroleum exploration north of the 62nd parallel:

> (a) In view of the general uncertainty involved in starting operations in a new area it is desirable that the factors known from geophysical surveys (should) indicate good prospects.
> (b) An area which is opened up for further investigations and exploratory drilling should contain different types of prospects, i.e. different types of traps. The traps may be of the same type, but should then be located in different geological strata. Should drilling prove one type of prospect uninteresting, then it will be possible to concentrate activities upon another type. This ensures a certain continuity and stability in operations. If all efforts are concentrated upon especially promising prospects as would be natural if limited venture capital is available for exploration, interest in the area may decline and may easily delay further exploration if the first wells drilled give negative results. This would be most undesirable from the exploration point of view . . .

(c) Another objective when selecting an exploration area is to obtain geological information which provides a basis for evaluating the total petroleum resources of the continental shelf . . . it is important that the geology within the areas selected is representative of greater parts of the shelf. The geological information obtained from drilling within these areas will then be useful in attempting to evaluate the resources on greater parts of the shelf.[52]

Developing country governments intead of adopting a passive role in relation to demarcation of areas for exploration and waiting for companies to indicate their preference, as has often been the case, can take the initiative in drawing up an overall strategy for allocation based on a broad assessment of its prospective areas. If this is not done, the objective of securing a rapid and thorough exploration of all its prospective areas may not be attained, as companies having selected a few areas in a country for exploration may withdraw from that country if no discoveries are made in those areas, leaving the rest of the prospective areas unexplored. An unplanned allocation based exclusively on company preferences may also prejudice the interest of securing an overall evaluation of the petroleum potential of a country's prospective areas.

Alternative procedures for allocation of areas and selection of companies

Where an allocation of an area for exploration has to be made to one of a number of competing applicants, alternative procedures may be adopted, such as an auction, or other forms of competitive bidding, or discretionary allocation. The relative advantages and disadvantages to a government of each of these procedures merit consideration.

An auction system such as is followed for granting of petroleum leases in the United States, and in Alberta and Saskatchewan in Canada, involves the simplest form of competitive bidding, where a license is awarded to the highest bidder of a cash payment (bonus). There can be schemes where instead of "bonus-bidding", bids may be invited from parties in terms of profit-sharing percentages, or of royalties that they are willing to pay. The relative merits of these different types of competitive bidding have been evaluated (in the context of US oil shale leasing policy), thus:

The least satisfactory method is profit sharing. Under this system the buyer bids a percentage of profits for the lease. The advantages of this system are limited. Barriers to entry are low because payment depends on a later profit . . . A further advantage is that windfalls are shared. Problems in using the method, however, are sufficient to exclude this method. The incentive to increase development and operating efficiency is low since a larger percentage of the gains must be paid to the lessor. Also deterring application of this scheme is the difficulty in administration. Because the cost is determinative of profits the Government would be required to have some degree of control over cost input. This would require excessive involvement in accounting and planning functions. Finally because this method selects the bidder who is willing to pass on the largest percentage of profits to the Government, the most efficient producer may not be selected . . . The remaining two schemes, bonuses and royalties, have significant advantages and disadvantages when considered alone. When used in an inter-related system, they can produce the important objectives sought by oil shale policy because the qualities of each system will offset some of the disadvantages of the other system.

Bonus bidding has several advantages that are incorporated into the combination system. The system is easy to administer because the bonus payments are set by the bid rather than being based on a complex formula. Also, the maximum incentive is provided since the company receives profits from an increase in efficiency under the bonus system . . . [Since royalties are] to consist of only a percentage of the increased profits . . . increased efficiency means greater return to the lessee . . . The Government is also protected by the sliding-scale percentage because windfall potential is shared by both parties to the lease.[53]

As between open bidding and submission of sealed bids, the advantage of calling for sealed bids is assessed thus:

Under oral bidding, if competition is limited, the bidding price may be low and the bids are vulnerable to collusion among the bidders in order to drive the . . . price down. Sealed bids contain an element of uncertainty due to the possibility of unknown competition, whereas under oral bidding all bids are known. The possibility of this unknown bid will force a serious bidder to align his bid more closely with fair market value.[54]

The arguments in favour of an auction system for allocation are: (a) it is more likely to ensure that the economic rent (the difference between total revenues and total cost, including the costs of management and an appropriate risk premium) or at least the greater part of such economic rent will be captured by the government, in the form of the bonus that will be bid; (b) it will tend to select the company which is the most efficient in the sense of having the lowest costs; (c) it provides an objective basis, thus eliminating the risk of allocations being made on the basis of irrelevant, or in cases illegitimate considerations, as would be the case if allocations were to be left to the untramelled discretion of the allocating agency.

The arguments adduced against an auction system are: (a) it could lead to allocation to a company which had bid a high bonus, only to pre-empt others from obtaining the area, and was itself not able or willing to undertake an effective exploration programme in the area; (b) the company making the highest bid may not be the one which is the most competent, or which possesses the greatest amount of financial or technological resources; (c) "experience elsewhere indicates . . . that, because of the gamble involved, the auction system usually discounts the true value of a prospective area";[55] (d) this system does not enable more complex criteria of selection reflecting policy objectives, such as preference in selection to national companies or companies committed to more active exploration programmes, to be applied.

In Britain, the issue of auction-versus-discretionary allocation was discussed in depth by the Public Accounts Committee of the House of Commons in the wake of the fourth round of licensing in 1971. In that round while the bulk of the blocks (267) had been allocated on a discretionary basis as had been done in previous rounds, fifteen blocks had been put up to auction, on an experimental basis "to test the market". The total amount bid for the fifteen auctioned blocks was £37 million, while the fixed initial payment obtained on account of the 267 blocks allocated on a discretionary basis amounted to £3 million. In response to the committee's probing questions as to why the auction method should not have been used in preference to the discretionary method for all the blocks, the principal arguments put forward by the Ministry were that under the auction system it would not be possible to ensure an appropriately rapid rate of exploration, nor would it have been possible to accord preference to British companies. When it was suggested to the Ministry that minimum work obligations could have been incorporated into the auction system, the Ministry representative explained the difficulties involved in doing so in the following terms:

> The exploration programmes which have been insisted on as part of the discretionary licensing system would be very difficult to apply in an auction, and absolutely impossible in the early days of exploration of a province, because not enough knowledge exists about the province to discuss and lay down an exploration programme or a work programme, such as is part of every license agreement under the discretionary system. It could not be done.
>
> . . . It would be possible to put certain conditions to the invitation to competitive bidding . . . one could say that one would have as a condition of the tenders that one hole should be drilled in each block . . . The important point to be taken about this is that the more you elaborate the conditions attaching to the auction type system the

nearer it comes to a discretionary type and the more liable you are to be brought into the area of judgments which are very difficult to administer if the purpose is that it should be on a clear-cut auctioning basis. To take the simplest case which I gave as my first example – one hole per block or per licensed area – that is simple but it is not necessarily right. It has not been the case that in all the discretionary licences given it has been one hole. There has been much discussion about the area, the type of work to be done, the quantity of seismic work to be done, whether there should be a hole or not, whether there should be more than one hole. All this depends on the nature of the block. I do not think that it would be practicable, if you are to make sense of an auction system, to leave too much discretion. You have got to keep it sharp, if you are going to have competitive bidding. The more conditions you attach to it the less liable you are to prove what you are really trying to prove with an auction, i.e., what is the market valuation of the territory.[56]

Most governments thus continue to favour the discretionary system. The discretion to be exercised, however, need not be untramelled, since certain objective criteria can and ought to be laid down to guide the selectors and elements of competition can be introduced into the selection process. Thus, for example, certain objective criteria are implicit in the legal provision which requires certain types of information to be submitted by applicants for production licences under the Norwegian system. Thus the Norwegian legislation requires applicants to produce, *inter alia*, the following information:

(Section 14) –

(c) . . . information as to form of organization, including information as to the relationship to the parent company and other integrated corporate structures that may be involved, petroleum production, refining and marketing possibilities and the company's own petroleum requirements.
(d) . . . details as to the financial structure of the applicant and its parent company including the annual reports for the last three years together with copies of the balance sheets and the profit and loss account for the same period for both the applicant and the parent, as well as information as to the manner in which the operations are to be financed.
(c) Information as to the applicant's previous experience in the exploration for and exploitation of petroleum.[57]

Similarly, certain objective criteria are implicit in the enumeration of considerations to be taken into account in the awarding of licences which are set out in the notice inviting applications for licences in the British sector of the North Sea, thus:

(3) the consideration (inter alia) the Secretary of State will have in mind in examining applications will be:—

(b) the extent to which the applicant will further the thorough and rapid exploration of the oil and gas resources of the United Kingdom Continental Shelf, particular attention being paid to the financial and technical ability of the applicant to carry out an acceptable work programme . . .
(c) exploration work already done by or on behalf of the applicant, which is relevant to the areas applied for;
(d) where the applicant already holds a production licence or licences, his overall performance to date;
(e) the extent of the contribution which the applicant has made or is planning to make to the economy of the United Kingdom including the strengthening of the United Kingdom balance of payments and the growth of industry and employment.[58]

The actual process of selection adopted in Britain, when awarding licences in the first round in 1964, as it has been described, indicates that an element of competition was involved in that process, thus:

Faced with the concentration of applications, the ministry made an initial provisional allocation for its own purposes. In the process of this allocation it excluded from consideration all applicants who could be excluded on the basis of established criteria. For example companies that did not have the requisite financial resources to carry out a drilling program but had applied only in the hope of selling their licence at a profit could be excluded at that point . . . To aid in the final allocation, each selected applicant was asked to submit a detailed statement of relative preferences as among different blocks and different combinations of blocks . . . More significant, however, was the ministry's request for a statement of each company's work program for the blocks for which the company was still running. It came to be known that the ministry expected much more active drilling programs in areas that were widely sought after than in less coveted areas . . . a "going price" came to be known for each area . . . This going price was denominated in such things as holes drilled and exploration work undertaken. He who was unwilling to pay the going price could not expect to be awarded a licence. The system could thus be characterised as a competitive bidding system in which the bid was the work program of the applicant . . . where an applicant's work program for a particular block seemed insufficient to the ministry, he was informed that unless he increased the extent of exploration and drilling activity he could not expect to receive a final allocation of that area. By means of this kind of direct negotiation, the ministry was able to introduce an element of competition into work programs.[59]

In practice, therefore, discretionary allocation systems can combine elements of competitive bidding and negotiations. Thus, by requiring the types of information that are required both under the Norwegian and the British systems, by enunciating factors to be taken into account in allocating areas, and further by stipulating "minimum terms and conditions" such as has been done by a number of countries,[60] when inviting applications, a framework can be established which reduces the risk of arbitrary allocation, by introducing certain objective criteria of selection and an element of competition. A good example of such a framework is the one elaborated and set out in the "consultative document" (reproduced in the Appendix to this chapter) issued by the UK Department of Energy, on the basis of which companies were invited to submit their bids for the Sixth Round of Licensing (May 1978). Thus in responding to such an invitation, the applicants not only signify acceptance of the "minimum terms", including participation provisions, but are called upon to provide evidence of their technical and financial competence, and to "bid" in terms of minimum expenditure and minimum work obligations they are willing to undertake for exploration. Upon receiving such "competitive bids", the government agency concerned can prepare a short list. In drawing up a short list, a "points system" may be followed which would award points not only to the bonus offered, or the minimum expenditure or work obligations, which a company offers to undertake, but its technical and financial competence, its past performance, its current commitments and resource position, the production currently available to it to meet the needs of its marketing outlets, and the potential markets that it is in a position to serve. Negotiations may then be undertaken with those companies which are on the short list and through this process a final selection may be made.

APPENDIX

SIXTH ROUND OF OFFSHORE PETROLEUM PRODUCTION LICENSING

CONSULTATIVE DOCUMENT (UK DEPARTMENT OF ENERGY — 1978)

Introduction

1   This document sets out the arrangements which the Secretary of State proposes for the Sixth Round of offshore licensing which is to take place this year.

Size of Round

2   The Round will be of about 40 blocks, with a wide geographical
spread. This Round is the next stage of the Government's strategy of
licensing smaller amounts of territory at more frequent intervals.

Awards of Licences

3   Applications will be invited on the basis that The British National Oil
Corporation (BNOC) will be a co-licensee with a full equity interest in
each licence of 51%, save that where the British Gas Corporation (BGC)
is also a co-licensee the combined interests of BNOC and BGC will be 51%.
The equity share for BNOC and/or BGC may be higher than 51% in cases
where the applicant accepts such an arrangement (see paragraph 9 below).
A licence will be issued to a successful applicant only upon:

(a) the conclusion of a joint operating agreement between the BNOC
and the applicant covering the licensed area in a form satisfactory
to the Secretary of State;

(b) the settlement between BNOC, the applicant and the Secretary of
State of a work programme for the initial and second terms of the
licence; and

(c) as appropriate, the continued satisfactory performance of the
applicant in accepting and implementing majority state participation in existing licences.

Guidelines will be published setting out matters which the Secretary of
State will expect to be included in the joint operating agreements. Proposals for these guidelines are set out in Appendix 1 which is divided into
two parts: Part A deals with matters other than the special features
referred to in paragraph 9(b)(i), (iii) and (iv) below; the special features
are dealt with in Part B. The Secretary of State will be prepared exceptionally to consider variations from the guidelines which he is satisfied
are reasonable and are consistent with the requirements of the national
interest.

Approval of Operators

4   In the Fifth Round Licences, where the Secretary of State has exercised the power given to him under the Petroleum (Production) Regulations 1976 to approve the appointment of operators, this approval has in
general (though not exclusively) been given for both exploration and
development. However, the scale of the resources of specialised manpower
and of technological capability that are required by an operator in the

exploration phase are very much less than those which are called for in the development of virtually all offshore fields. And the total of the necessary resources as a whole is not readily capable of expansion. There would therefore be advantage if the decision as to who should be operator for the development stage is deferred until such time as a discovery is made and development is contemplated, when the decision can be made in the light of then current information about matters such as the availability of resources. In future therefore the Secretary of State will initially approve the operator for the exploration phase only.

5   The Secretary of State will expect BNOC to be operator for the exploration phase in six blocks, the selection of which will be made after applications have been received.

Arrangements for Applications

6   The following changes are proposed:

(a)   the level of the application fee for offshore production licences is to be increased; and

(b)   each application is to be accompanied by supporting financial information in the form of annual audited accounts of the applicant, and any company that controls the applicant.

Conditions of Licences

7   The conditions of licences will, in general, be left as set out in the Petroleum (Production) Regulations 1976, including the controls over development programmes and the rate of production. These controls are particularly important to the Department's objective of obtaining an orderly programme of development and production, and so ensuring that the national interest is properly secured in these matters. The only change proposed to the licence conditions is that the arrangements for the approval of operators are to be made clearer, in the light of the considerations mentioned at paragraph 4 above.

Financial Terms

8   The proposed terms are set out in Appendix 2. These include royalty at 12½% on tax value.

Factors Governing the Assessment of Applications

9   Licences are to be issued solely at the Secretary of State's discretion. The criteria proposed for the Sixth Round are attached at Appendix 3.

These are in the main those which applied for the Fifth Round. In addition however the following new factors will also be taken into account:

(a) the applicant's record in providing training for work on offshore installations; and

(b) whether the applicant is prepared:

   (i) to meet BNOC's share of exploration and appraisal costs under the licence and, if so, what proportion thereof;

   (ii) to accept an equity interest for BNOC in the licence of more than 51% and, if so, what percentage;

   (iii) to grant to BNOC the option to purchase from the applicant at market price the applicant's share of oil and natural gas liquids produced under the licence and, if so, what proportion thereof; and

   (iv) to grant to BNOC the option to sell to the applicant at market price BNOC's share of oil and natural gas liquids produced under the licence and, if so, what proportion thereof.

References in (b) above to BNOC where appropriate include BGC. Guidelines as to the basis upon which the Secretary of State will expect the matters referred to in (i), (iii) and (iv) above to be included in the relevant joint operating agreements are set out in part B of Appendix 1. Applicants will be required to include their offers in respect of the matters referred to in (b) above as attachments to their applications.

10   The Secretary of State will consider applications only for such blocks as are specified by block number in each application.

11   There will be no restriction of applicants on grounds of nationality.

APPENDIX 1

PROPOSED GUIDELINES AS TO MATTERS TO BE INCLUDED IN JOINT OPERATING AGREEMENTS

PART A

The Operator

1   Initially, the appointment of the operator shall require the agreement of all the prospective licensees; but thereafter such appointment will be by vote of the Operating Committee. Subject to giving appropriate notice,

the operator is to have the right to resign at any time and the operating committee is to have the right to remove the operator at any time. All appointments of the operator are subject to the approval of the Secretary of State.

The Operating Committee

2   An operating committee shall be established to exercise overall supervision and control of all operations conducted under the licence. Such committee shall consist of one representative appointed by each licensee. Each licensee shall have a voting interest equal to its interest in the licence. Unless otherwise agreed by BNOC all meetings of the Operating Committee and any related Committees shall be held in Glasgow.

Decisions of the Operating Committee

3   Decisions of the operating committee shall be made by the affirmative vote of the licensees. The percentage of votes required for decisions shall be agreed upon by all the prospective licensees but must be within the range 52% to 75% provided that BNOC is not to be able, solely by the use of its majority vote, to commit other members of the operating committee to expenditure.

Surrender of Licence Area

4   In respect of any decision of the operating committee relating to compulsory relinquishment of part of the licence area, BNOC will not use its vote to frustrate the unanimous wishes of the other licensees desiring to continue the licence. The agreement is not to contain assurances or undertakings between the co-licensees as to future involvement in surrendered acreage or unlicensed blocks.

Work Programmes and Budgets

5   The operating committee shall decide upon the programmes of work to be carried out under the licence and the budgets therefor. In the case of any exploration or appraisal programme and budget which is approved by the operating committee, all parties shall be bound to participate and accordingly there will be no right of non-consent. In the case of any development programme and budget which is approved by the operating committee, each party shall have the right within a reasonable period to decide whether or not to participate. Subject to paragraph 6 below, each party will be liable for its share of costs and obligations.

Sole Risk

6   In the event of the operating committee failing to approve a proposed programme of work, or a party deciding not to participate in a development programme budget approved by the operating committee, any party shall be free to carry out such work at sole risk provided it does not interfere with joint operations and provided the sole risk party agrees to indemnify the non-sole risk parties against all claims and proceedings brought by any third party arising out of the sole risk operations. The sole risk arrangements shall enable non-sole risk parties to re-join the venture at any time up to the point at which the Secretary of State consents to the commencement of any development works relevant to the production of petroleum in respect of that venture. In the event of BNOC deciding not to take part in any development, it shall nevertheless be entitled to be represented at all meetings at which that development is discussed and to receive all data and information relating to that development. BNOC will not however be entitled in respect of its equity interest to a share in the petroleum produced from that development, nor to a share in the ownership of assets relevant to that development, nor to a vote on matters arising in respect of that development.

Detailed Financial Arrangements

7   The operating agreement is to include arrangements for the detailed financial management of the licence activities, such as the dates on which the licensees' contributions are to become due, provisions for default in such contributions and accounting and auditing procedures.

Assignments

8   In the event of the Secretary of State directing BNOC or BGC to assign its interest in the licence to the other corporation or any of its subsidiaries, BNOC and BGC will be entitled so to assign its interest. BNOC is to have the first option as to all or any part of any assignment of interest (other than to an affiliate) proposed by any of its co-licensees. All assignments are subject to the approval of the Secretary of State.

Public Announcements

9   Where BNOC is not the operator, BNOC is to have the right to assume joint responsibility with the operator for all public announcements and statements concerning the licence activities.

BNOC's Involvement in Joint Operations

10 Where BNOC is not the operator, BNOC is to have the right, with the concurrence of the operator:

(a) to second personnel to the operator for work on the joint operations; and

(b) itself to undertake on behalf of the operator any tasks which have been approved by the Operating Committee.

PART B

Carried Interest for BNOC

11 Where BNOC's co-licensees are to meet all or part of its share of exploration and appraisal costs (see paragraph 9(b)(i) of the main part of this document), the following arrangements are to apply:

(a) BNOC's rights as a full equity licensee are not to be diminished in any way by reason of such carried interest;

(b) BNOC will not pay for any costs or be liable for any obligations which would otherwise be attributable to the share for which it is being carried, except and to the extent that it elects to participate in any development programme. If BNOC does so elect, it will reimburse the share of costs incurred in relation to the particular discovery which it would have paid if it had not been carried and will pay its share of the costs of the development programme (but not of any other licence activities); and

(c) BNOC will reimburse its co-licensees at the time when it elects to join in a development.

BNOC's Options to Purchase and/or Sell Oil and Natural Gas Liquids

12 Where BNOC is to have the option to purchase and/or sell oil and natural gas liquids (see paragraph 9(b)(iii) and (iv) of the main part of this document) the following arrangements are to apply:

(a) BNOC will give at least six months written notice of its intention to purchase or sell (as the case may be); and

(b) the price payable or receivable by BNOC shall be the market price, which shall be ascertained by reference to agreed principles, and shall be capable of settlement by an expert in the event of a dispute.

SIXTH ROUND OF OFFSHORE LICENSING —
FINANCIAL TERMS

The proposals are:

Royalty	12½% on tax value
Application fee	£1,250
Periodic payments for licences	
— for the 1st period of 4 years	
one payment of	£100 per sq km
— for the following period of 3 years	
one payment of	£150 per sq km
— for the 8th year	£250 per sq km
rising on an annual incremental	
scale of	£250 per sq km/pa
— to a maximum of	£3,750 per sq km/pa

PROPOSED CRITERIA FOR SIXTH ROUND

Applicants will be judged against the background of the continuing need for expeditious, thorough and efficient exploration to identify oil and gas resources of the UK Continental Shelf, and the following factors will be particularly borne in mind when examining applications:

(a) technical competence to undertake a programme of exploration and production;

(b) capability to produce funds commensurate with work programme obligations in respect of initial exploration and the extent of access to adequate funds in the event of a commercial discovery being made;

(c) where the applicant already holds or has held a licence, his overall performance to date in meeting licence obligations;

(d) exploration already done by or on behalf of the applicant which is relevant to the areas applied for;

(e) the extent of the contribution which the applicant has made or is planning to make to the economy of the UK, including the strengthening of the UK balance of payments and the growth of industry and employment;

(f) where a body incorporated in a country outside the United Kingdom applies for a licence or holds a controlling interest in the applicant, how far equitable treatment is afforded in such other country;

(g) the degree to which the applicant, or any licensee in whom he has a controlling interest, or any licensee who has a controlling interest in the applicant, has demonstrated his agreement to majority State participation in any discovery made under existing licences and where appropriate his effective implementation of such arrangements;

(h) whether the applicant subscribes to the Memorandum of Understanding agreed by the Secretary of State and United Kingdom Offshore Operators Association to ensure that full and fair opportunity is provided to UK industry to compete for orders of goods and services. Where the applicant is or has been a licensee, his past performance in providing full and fair opportunity to UK industry will be taken fully into account;

(i) whether the applicant subscribes to the Memorandum of Understanding to grant reasonable access to representatives of independent trade unions to his offshore installations. Where the applicant is an existing licensee his performance in providing such access will be taken fully into account;

(j) the applicant's record in respect of training for employment on offshore installations;

(k) whether the applicant is prepared to meet BNOC's or BGC's share of exploration and appraisal costs under the licence, and if so what proportion thereof;

(l) whether the applicant is prepared to accept an equity interest for BNOC or BGC in the licence of more than 51%, and if so what percentage;

(m) whether the applicant is prepared to grant to BNOC the option to purchase from the applicant at market price, the applicant's share of oil and natural gas liquids produced under the licence, and if so what proportion; and

(n) whether the applicant is prepared to grant to BNOC the option to sell to the applicant at market price BNOC's share of oil and natural gas liquids produced under the licence, and if so what proportion.

NOTES

1. See Cattan, *The Evolution of Oil Concessions in the Middle East and North Africa*, pp. 1-25.
2. David N. Smith, "Mining Resources of the Third World: From Concession Agreements to Service Contracts", 1973 American Society of International Law, *Proceedings of the 67th Annual General Meeting*, p. 228.
3. David N. Smith and Louis T. Wells, *Negotiating Third World Mineral Agreements*, p. 12.
4. R. Fabrikant, *Legal Aspects of Productions Sharing Contracts in the Indonesian Petroleum Industry*, p. 104, quoting Louis Wells: "A concession contract is the product of a bargaining process, reflecting the strengths and weaknesses of the two parties and their bargaining skills. But the relative positions of the parties change with time . . ." in *The Evolution of Concession Agreements*, Economic Development Report No. 117, Development Advisory Service Conference, Sorrento, 1971.
5. G. O. Gutman, "Objectives, Strategy and Tactics in Negotiations for Mining Projects", in *Foreign Investment, International Law and National Investment*, edited by J. G. Zorn and P. Bayne, p. 99. See also T. H. Moran, *Multinational Corporations and Political Dependence*, pp. 159-61, where the shifts in the bargaining power of the parties to a mining venture are described thus: "The foreign investor starts from a position of monopoly control over the capacity to create a working operation out of a potential ore-body − a monopoly control that only a few alternative competitors could supply at a broadly similar price. There is always a great deal of uncertainty about whether the investment can be made into a success and what the final costs of production and operation will be. The government would like to see its natural resource potential become a source of revenue and employment, but the government cannot itself supply the services needed from the foreign investor and is even less qualified than the investor to evaluate the risk and uncertainty involved . . . The conditions under which a foreign company will agree to invest must initially reflect both his monopoly control of skills and his heavy discounting for risk and uncertainty. The host government may want to get as much as possible from the new venture. But the strength of the bargaining is on the side of the foreign investor, and the terms of the initial concession are going to be heavily weighted in his favor . . . But once the investment is made and the operation is a success, the whole atmosphere that surrounds the foreign (investor)–host relationship changes. The old doubts are forgotten, and the terms of the concession no longer correspond to the 'realities' of the situation. A gamble with large risks has been won, and the host government is unlikely to want to keep paying premiums that reflect those risks for long . . . [At the same time success reduces uncertainty for subsequent investors and] the government can drive a tougher bargain with later entrants, and this in turn increases the leverage in demanding revision of the original concessions to put them in line with later agreements . . . In short, with the reduction of uncertainty the bargaining strength inevitably shifts from the foreigners toward the host government."
6. R. Fabrikant, *Legal Aspects*, op. cit., p. 105. Also Mikesell, *Foreign Investment in Petroleum . . .*, op. cit., pp. 38-39.
7. Fabrikant, op. cit., p. 105.
8. R. Brown and M. Faber, *Some Policy and Legal Issues affecting Mining Legislation and Agreements in African Countries*, p. 10.
9. Fabrikant, op. cit., p. 110.
10. See Kuwait − Concession Agreement with the Arabian Oil Company of 5 July 1958, published in *Selected Documents of the International Petroleum Industry*

(*Iraq and Kuwait Pre-1966*), OPEC, p. 153 at p. 158.

11. See Greece–Texaco Agreement of 28 September 1968, published in *Selected Documents of the International Petroleum Industry*, 1968, OPEC, p. 289 at pp. 291–296, and Angola-Sun Oil Group Agreement of 21 March 1974, *Petroleum Legislation* (South and Central Africa), Supp. XXXVIII.
12. This account was obtained from interviews with representatives of US oil companies, July 1977.
13. This point was made by almost all the American companies which were interviewed, July 1977.
14. Moran, op. cit., p. 254.
15. Ibid., pp. 163–164.
16. Fabrikant, op. cit., p. 169.
17. Smith and Wells, op. cit., p. 166.
18. Ibid.,
19. B. F. Grossling, *Latin America's Petroleum Prospects in the Energy Crisis*, U.S. Geological Survey Bulletin 1411, p. 26.
20. Interviews with US oil companies, July 1977.
21. Ibid.,
22. R. Vedavalli, *Private Foreign Investment and Economic Development – A Case Study of Petroleum in India*, p. 141.
23. Interviews with US oil companies, July 1977.
24. Moran, op. cit., p. 255.
25. K. W. Dam, *Oil Resources, Who Gets What How?*, pp. 124–130.
26. T. H. Moran, "Transnational Strategies of Protection and Defense by Multinational Corporations: Spreading the Risk and Raising Cost of Nationalization on Natural Resources", *International Organization*, Vol. 27, p. 273. Also Moran, op. cit., p. 255.
27. Interviews with US companies, July 1977.
28. T. Waelde, "Lifting the Veil from Transnational Mineral Contracts: a Review of Recent Literature", *Natural Resources Forum*, 1977, Vol. I, p. 277 at p. 282.
29. J. Favre, "Exploration in Petroleum Deficit Countries", Paper read at U.N. Meeting on Petroleum Co-operation among Developing Countries, Geneva, 1975 (ESA/NRET/AC. 10/19).
30. L. Howell and M. Morrow, *Asia, Oil Politics and the Energy Crisis*, p. 63.
31. J. Favre, op. cit.,
32. See Franco-Algerian Oil Accord of 1960, and the Iraq-Soviet Union Agreement concerning Economic and Technical Co-operation in the Development of National Oil Production, *Petroleum Legislation* (Middle East), Suppl. XXIV.
33. *Petroleum Co-operation among Developing Countries*, U.N. STA/ESA/57, pp. 115–116.
34. AGIP Mineraria (ENI subsidiary)–NIOC Joint Venture Agreement of 3 August 1957; ERAP–NIOC Service Contract for Technical, Financial and Commercial Services of 12 December 1966, *Petroleum Legislation* (Middle East), Supp. XV.
35. H. Zakariya, "New Directions in the Search for and Development of Petroleum Resources in Developing Countries", *Vanderbilt Journal of Transnational Law*, 1976, Vol. 9, p. 545 at p. 563.
36. OAPEC, Annual Report, 1975.
37. *Petroleum Co-operation*, op. cit., pp. 124–127.
38. B. F. Grossling, op. cit., p. 11.
39. J. H. Favre, op. cit.,
40. Interview with representatives of the Mexican Petroleum Institute, August 1976, Statoil of Norway, March 1977, and US oil companies, July 1977.
41. Fabrikant, op. cit., p. 26.
42. The importance of independent evaluation was repeatedly emphasised in the

course of interviews by representatives of Statoil and the Norwegian Petroleum Directorate, March 1977.

43. Fabrikant, op. cit., p. 26.
44. Interview with representative of Norwegian Petroleum Directorate, March 1977.
45. Northcutt Ely, "Policy considerations in the development of mineral laws", *Natural Resources Lawyer*, 1970, Vol. 3, p. 281 at p. 287.
46. Ibid.,
47. G. W. Stocking, *Middle East Oil*, p. 166; Lenczowski, *Oil and State in the Middle East*, p. 91.
48. Dam, op. cit., p. 24.
49. Ibid., p. 26.
50. Ibid., p. 28.
51. *Report No. 30 to the Norwegian Storting* on Operations on the Norwegian Continental Shelf, 1973-74, pp. 63-64.
52. *Report No. 91 to the Norwegian Storting* on Petroleum Exploration North of 62°N (1975-76), p. 34.
53. J. Denny Moffett, "Federal Oil Shale Policy: An Analysis of Development Alternatives", *Houston Law Review*, 1976, Vol. 13, p. 701 at pp. 719-721.
54. Ibid., p. 718.
55. *Ireland Exclusive Offshore Licensing Terms*, Department of Industry and Commerce, Government of Ireland, 1975, p. 20.
56. *First Report on North Sea Oil and Gas*, Public Accounts Committee, House of Commons, Session 1972-73, p. 3 at p. 66; for a critique of the Ministry's arguments, see K. W. Dam, "The Evolution of North Sea Licensing Policy in Britain and Norway", in *The Journal of Law and Economics*, 1974, Vol. XVII, p. 213 at pp. 215-221.
57. Norwegian Royal Decree of 8 December 1972 relating to Exploration for and Exploitation of Petroleum in the Sea-bed and sub-strata of the Norwegian Continental Shelf.
58. *First Report*, Public Accounts Committee, cited above, p. 47.
59. Dam, *Oil Resources*, op. cit., pp. 27-28.
60. See *Minimum Terms for Service Contracts*, issued by CVP, the Venezuelan National Oil Company, published in *Selected Documents of the International Petroleum Industry*, 1968, p. 381; *General Terms and Conditions for Offshore Oil and Gas Bids*, issued by Myanma Oil Corporation, the Burmese National Oil Company, in 1973, published in *Petroleum Legislation* (Asia and Australasia), Supp. XXXIX, A-O.

CHAPTER IV

The Legal Framework and the Choice of Legal Mechanisms

Having formulated its broad strategy, a government has to establish a legal framework or, where one exists, to revise it, so as to incorporate in it appropriate legal mechanisms, to secure its strategic objectives. It has to determine the matters which are to be dealt with by general legislation and those which are to be settled by negotiation and embodied in contractual documents. It has also to choose a type of petroleum development arrangement from among a wide range of different forms of arrangement which have been developed in different countries.

General legislation or individually negotiated agreements

The question arises as to whether the legal framework in which companies are to be involved in petroleum development should be defined by general legislation, or by individually negotiated (*ad hoc*) agreements or a combination of both. The practice of countries reveals different approaches which fall broadly into three categories:

1. *General legislation system*
Under this system, legislation fixes in advance conditions under which rights to explore for and/or exploit petroleum resources may be granted under standard form licences or leases; royalty taxes and other payments to be made are also determined by legislation. The countries in which this system is in force include the United States, Canada, Australia and most of the EEC countries.[1] A number of countries which might, at first sight, appear to fall in this category, such as Britain, New Zealand and Norway, cannot, on closer examination, be properly so classified since, although the broad legal framework in their case is defined by general legislation, the legislation contains provisions which leave a number of important matters, such as the extent of the work obligation during the period of exploration, to be settled by negotiation.

2. *Individually negotiated (*ad hoc*) agreements*

Under this system, there is no general system of legislation, or the legislation is of a very general nature, which in effect leaves it to the government to grant rights to explore for and/or to exploit petroleum resources on the basis of individually negotiated agreements. This was the case with the early concessions granted in most of the traditional producing countries, such as Iran, Iraq and Saudi Arabia. More recent examples are provided by countries which have legislation which lays down very broad general principles but leave terms and conditions on which companies may be employed for petroleum development to be determined by negotiation. Thus, for example, the basic Law No. 44 of 1960 in Indonesia, while laying down that "the mining of oil and gas should only be undertaken by the state and that mining undertakings of mineral oil and gas are exclusively carried out by state enterprises" (Article 3), leaves scope for employment of foreign companies as "contractors" on the basis of specifically negotiated agreements. The Bangladesh Petroleum Act of 1974 similarly lays down the broad principle that petroleum operations are to be carried out exclusively by the government or state-owned companies, while leaving it to the governmental agency concerned to employ foreign oil companies as "contractors" on terms to be negotiated with them.

3. *Hybrid system – general legislation and individually negotiated*
   agreements

Under this system, general legislation lays down certain fundamental provisions and stipulates certain minimum standards and conditions which must be satisfied by applicants for the grant of the right to explore for and/or exploit petroleum resources but provides for certain important terms and conditions to be settled by negotiation. Among the countries which fall into this category are Britain, India, Malaysia, the Netherlands, New Zealand, Norway and Trinidad and Tobago.

Thus, the Norwegian legislation[2] elaborates a framework in which companies may obtain a non-exclusive reconnaissance licence (Chapter 2) or a production licence (Chapter 3). Among the matters fixed by legislation are the initial duration of the production licence – 6 years (Section 15) and the condition that no production licence may be granted "until the Ministry, after consulting with the applicant, has approved a work programme" (Section 17). It further prescribes a number of conditions which must be accepted by applicants: that the Ministry shall fix the magnitude of the amount which shall be paid to the Ministry in the event that the applicant fails to execute the agreed work programme; that only upon fulfilment of the work programme can the company demand that the licence remain valid for a period of 30 years for an area of up to one-half of the original licence area (Sections 20, 22); that upon the adoption of a budget and work programme for the coming years, these shall be forwarded to the Ministry; that there would be a service or rental fee, the rate for which would progressively increase from year to year (Section 25); that there would be a sliding scale royalty increasing with the volume of production from a field (Section 26); that the licencee shall carry out his operation from a base in Norway and that his organisation in Norway should be sufficient to direct this activity and to make all decisions as to the activity and that the licencee shall use Norwegian goods and services in his operation as far as they are competitive in service, delivery schedule, and price (Section 54).

While all of these matters are laid down in the general legislation, a number of critical issues, such as the size of the exploration programme (which must be approved as a condition of the grant of a production licence) and the extent of state participation, the latter in the context of a provision that the Ministry may as a condition "for the granting of a production licence require state participation",[3] are matters which involve intensive negotiations. Indeed, the Norwegian experience is that in the early years the most difficult issue to negotiate was the size of the exploration programme, whereas now, after substantial discoveries have been made, the hardest negotiations relate to the extent of state participation.[4] State participation, in effect, involves negotiating a joint venture agreement (with a specially devised "carried interest formula" which will be discussed below). A number of important issues which have to be negotiated in this context are: the respective percentage interests to be held by each party, the management structure and control of operations, and the conditions under which the obligation to invest in development would arise.

In evaluating the alternative approaches outlined above, the first two should not be viewed in "either/or" terms, since it is evident that a hybrid system which combines elements of both is increasingly being adopted by countries. Further, in comparing these approaches, both legislation and agreements should be viewed as legal mechanisms which are to be used to secure specific objective and interests.

4. Relative merits of the general legislative system
The principal advantage from the point of view of the state, in adopting a general legislation system, is that the terms, in particular fiscal terms fixed by legislation, can be varied by subsequent legislation. The individually negotiated agreement, in contrast, is seen as an instrument which

"freezes" the terms, including the financial terms fixed by contract.

There is increasing recognition that petroleum development arrangements, like other mineral development ventures, being of relatively long duration, should incorporate mechanisms which would allow for adjustment of terms, and in particular the financial terms, to be made in response to changed conditions. That this problem antedates OPEC's recognition of it in Resolution XVI. 90 in 1968,[5] and the more recent pressures for change, is evident from an oil company memorandum prepared as early as 1946:[6]

A concession based on legal rights must provide a fair participation by both company and government in the development of any natural resources. If a concession proves *more productive than originally conceived*, the advantages to the government as a whole should be fairly weighed. If the additions to the economy of a country are not in fair proportion, governments will seek to impose added concession burdens. The art of maintaining a concession in force is the art of negotiating and compromising when such conditions arise. [Italic wording added.]

The need for devising mechanisms which would enable adjustment to be made of terms under which petroleum development was being carried out, and of the relevance of a general legislation approach, is recognised in a US State Department Policy paper on Middle East oil written in 1950:[7]

Experience in other parts of the world indicates that this movement towards increased national benefits and increased national control of oil operations is ubiquitous. If progress towards these national goals is obstructed by foreign companies, the process appears to become more accelerated or explosive. Therefore it is recommended (*inter alia*) that (a) The United States government should urge oil companies to include in concession contracts provisions calling for automatic re-negotiation of financial clauses every five to ten years. (b) The United States government should consider encouraging the adoption of general legislation rather than negotiated contracts.

The advantage of a general legislation approach in adapting terms and conditions in response to changing conditions, has been emphasised in a study of Canadian petroleum legislation, in which it is observed that:[8]

The oil-producing jurisdictions in Canada have developed a legal mechanism for controlling oil agreements, so that the relation between the government and the oil company can be made responsive to changes which the public interest dictates. This mechanism is a clause in oil agreements requiring the oil company to accept as binding all legislative and regulatory changes which may be enacted or promulgated from time to time in the future.

This mechanism enabled Alberta in the early sixties to respond to changes in the environment in the oil industry by resorting to legislation to alter existing petroleum development arrangements. The Alberta Mines and Minerals Act was amended in 1962. The term of renewal of existing leases was changed to 10 years (in place of 21 years) and drilling commitments were to be enforced over 5 years (instead of 1 year). It was also provided that existing 21-year leases would be brought to an end after 10 years if the lessee had not drilled and gained production or unitised the lease.

More recently in response to the 1973 price rises, both Britain and Norway enacted legislation introducing new financial provisions which would enable the government to appropriate a substantial part of the "excess" or "windfall" profit accruing to the companies.[8] Indonesia, where the terms are fixed by the contractual provisions embodied in production-sharing agreements, responded to the 1973 price increases by seeking alteration in the contractual terms, so that the existing cost-recovery allowance of up to 40 per cent of the crude oil production would be reduced to 25 per cent, and the ratio in which the remaining production would be shared between the government and the company would be alterated from 60/40 to 85/15. While companies protested the changes in each of the above cases, the legislative changes were more readily acquiesced in, while the reaction was strongest in Indonesia, where the companies concerted their opposition and refrained from undertaking any new exploration until the proposed amendments were substantially modified.[9]

Companies, while conceding that adjustment of terms may be justified in certain situations, which were marked by significantly changed conditions, are averse to general legislation regimes, under which the government retains the power unilaterally to alter the terms or to increase the financial burden on the company. They, therefore, express a marked preference for individually negotiated contracts, by which the terms can be frozen. In some of the agreements entered into in the fifties, as in the NIOC-AGIP joint venture agreement of 1957, a clause was incorporated in the following terms:[10]

> No general or special measure, legislative or administrative, or any other act of this kind emanating from the Iranian government, can annul this Agreement, amend or change its provisions, prevent or hold up the necessary and effective execution of these stipulations . . . The abrogation, amendment or modificationof the Agreement shall not take place without the unanimous consent of the parties. In case there is a difference between the provisions of the Agreement and existing laws, the provisions of this Agreement shall remain valid.

It remains questionable whether in law such a clause can nullify the impact of legislation which has the effect of altering the contractual terms embodied in an agreement.[11] Companies, however, tend to prefer individual agreements, for it has been observed that,[12]

> unsure whether the political process in the host country is such that the general laws will develop in reasonable ways, investors turn to agreements where terms will be fixed over a long time period.

Thus, in Trinidad and Tobago, when the Petroleum Act, 1969, introduced a new fiscal regime, which would increase government take and leave open the possibility of further increases by subsequent legislation, the companies prevailed over the government to enter into "production sharing agreements", under which the respective shares of the government and the company would remain frozen.[13]

It is no doubt technically possible to incorporate into individually negotiated agreements an elaborate re-negotiation clause, or a mechanism which in substance would operate in the way that the British Petroleum Revenue Tax operates, whereby when the rate of return on capital expenditure, measured on the basis of historical cost, exceeds a certain limit, a substantial part of the surplus profit would be appropriated by the government. In practice, however, companies remain strongly opposed to the incorporation of re-negotiation clauses, or mechanisms of the type indicated above.[14]

The insistence by some governments on the incorporation of some mechanism providing for re-negotiation of terms, has led to the inclusion in some agreements of a re-negotiation clause formulated in very general terms, thus:

(a) If, as a result of changes in the terms of concessions in existence on the Effective Date, or as a result of the terms of concessions granted thereafter, an increase to governments in the Middle East should come generally to be received by them, the Company will consult with the Emir, whether in the light of all the relevant circumstances, including the conditions in which operations are carried out, and taking into account all payments made, any alteration to the terms of the Agreement would be equitable to the parties (Article 27 — Kuwait-Shell Agreement, 15 January 1961).

(b) In case of changes taking place in the price structure of Middle East crude oil or in the Iranian taxation regulation which may affect the provisions of section 2 of the Article, the Parties to the Agreement shall meet in order to settle on new arrangements. Such arrangements shall ensure that neither Party shall profit at the expense of the other when compared with the present arrangements (Article 22 — NIOC

(Iran)-Tidewater Oil Co (and Others) Joint Structure Agreement, 16 January 1965).

(c) If, in the future, arrangements are made between the Government of Abu Dhabi, or of any other state in the Middle East, or the agent of any such government and the Companies or any other company or companies operating in the petroleum industry, as a result of which an increase in benefits should accrue generally to all such governments as aforesaid, then the Ruler and the Companies shall review and discuss the changed circumstances within the petroleum industry, taking into consideration operating conditions within Abu Dhabi in order to decide whether any alteration to the terms of the Agreement would be equitable to such parties (Article 37 — Abu Dhabi-Mitsubishi Agreement, 14 May 1968).

The decided reluctance of companies to incorporate sophisticated renegotiation clauses provides the strongest argument for fixation of terms, and in particular financial terms, by general legislation thus reserving the power to the government to make suitable alterations by subsequent legislation. The efficacy of this instrument is, however, contingent on its judicious use, for it has been rightly observed that:

the legal mechanism must not be given an exaggerated role in terms of sovereignty over oil resources. Legal regulations do not exist in a vacuum. For a government to reserve the legal power to modify the terms of oil agreements means little if the use of the power impairs the economic viability of the petroleum industry. In the end, the worth of the legal mechanism must be measured in terms of the willingness of foreign investors to participate in the petroleum industry on conditions acceptable to an informed and responsible government.[15]

A further advantage of the general legislation approach is that it enables certain general policy objectives to be written into the legal framework. Thus, for example, the Iranian Petroleum Act of 1957, laid down that at least one-third of the total exploitable territory was to be conserved at all times, and the Venezuelan Hydrocarbons Law of 1967 that "the terms and conditions stipulated in each contract must be more favourable to the owner [the state] than those provided under concessions".[16] Also, by general legislation, minimum standards and basic conditions can be laid down for grant of rights of exploration and/or exploitation of petroleum resources. Among the matters in respect of which standards and conditions can be laid down are: Information which a company must furnish regarding its financial and technical competence and experience;[17] the requirement that the company must incorporate a subsidiary in the country in which it will conduct operations;[18] maximum size of area which can be awarded to a single applicant;[19] maximum duration of the agreement;[20] relinquishment/area reduction requirement;[21] requirement that the company must commit itself to a minimum work and a minimum expenditure during the period of exploration;[22] information to be submitted during each phase of petroleum operations;[23] and safety requirements.[24] To leave these matters entirely to administrative discretion would expose the administrative agency concerned to great pressure from the companies, and deny to it the greater knowledge (and the confidence resulting from it), which it can acquire through undertaking the extensive preparatory work which must necessarily precede the formulation of a reasonably comprehensive piece of legislation.

To the legitimate criticism that to write in too much into the law could lead to lack of flexibility, so that standards may be set which are too high and would, therefore, deter companies from coming forward, or are set too low, and thus would prejudice the interests of the state, the answer must be that flexibility can be introduced in a number of ways. The most effective technique for introducing flexibility is to leave room for negotiation on a number of matters, where some variation should be reasonably expected having regard to such "objective" differences as location, geological and geophysical features, history of past exploration, etc. Among the matters which could be left to be settled through negotiation are: the extent of the minimum work and the minimum expenditure obligations during exploration, the extent of state participation (upon discovery), and the quantum of signature and production bonuses. The last, however, could also be left to be determined through competitive bidding. Another technique for introducing flexibility is while setting standards, to give to the administrative agency concerned power to grant exemption in special circumstances.

Thus, a strong case can be made out in favour of the hybrid system which is increasingly being adopted by governments. The hybrid system, by setting minimum standards and conditions, not only provides a measure of protection for government negotiators from pressures and inducements of the companies to concede to them terms more favourable than the "minimum", but also strengthens the government's negotiating position. The government starts off with the advantage that the company must accept the statutory minimum. It can thus invite companies to submit competitive bids offering terms better than the minimum. The government can seek improvement of terms through negotiations. In cases where the government is empowered to grant certain exemptions, it can trade off such exemptions against valuable concessions from the company. It has thus been observed that the interaction of statutory and contractual terms provides a framework to play off such inter-relationships and thus:

> can be an important bargaining asset for a host country; having established a firm and precise mining code with seemingly little room for exemptions, the host country can therefore sell exemptions from the code for discreet concessions [from the company]. Alternatively, the mining code may state minimum standards, leaving room for further bargaining defining the details of a project.[24]

In systems based on general legislation, provision is made in a number of cases for rights in respect of different stages of petroleum operations (reconnaissance, exploration (drilling), production) to be granted by separate documents: this too can confer a certain advantage on the government. The process may be split up into two or three stages. In the three-stage system, the first stage may be limited only to a reconnaissance permit (non-exclusive), the second stage to an exploration licence, and the third stage to a production licence or petroleum lease which is granted in the event of a commercial discovery being made. Under a two-stage system, the first stage is covered by an exploration permit or licence and the second stage, that is, after a commercial discovery is made, is covered by a production licence. Thus important advantages can be secured by a government if, for example, the grant of an exploration licence is made contingent on the submission of a satisfactory work programme and fulfilment of other conditions and the grant of a production licence, conditional on the making of a commercial discovery.

Some who favour a three-stage system, the first stage of which consists of a non-exclusive reconnaissance licence, argue that such a system enables vital correlation surveys to be carried out in respect of geological inspection and sampling and also geophysical surveys with the maximum flexibility and feedback of technical information to the government at an early stage. The argument against issuance of such non-exclusive reconnaissance licences over wide areas is that it confers an initial advantage on the companies, since in general their capabilities in making assessments and evaluations on the basis of the data collected by them are likely to exceed those of governments, particularly in developing countries. A more advantageous alternative for governments, where reconnaissance surveys are required to be carried out over large areas, is to have such reconnaissance surveys carried out through specialised contractors and to have the data so gathered independently evaluated with the help, if necessary, of consultants. This would enable the government to start with an initial advantage, as in that case only the government would have the total picture and the companies would have to purchase some of this information, and their knowledge would be limited to the specific areas in respect of which they may acquire rights. The trend is increasingly towards a two-stage system as, for example, is evident from the new legislation in 1967 which introduced a two-stage system in Australia, where previously a three-stage system had been in force.[25]

5. Petroleum agreements – the range of alternatives and the criteria for choice

In systems other than those in which relations between governments and companies are exclusively regulated by general legislation, a choice has to be made of a particular type of petroleum agreement from among a range of alternatives. New forms of agreements have been developed to replace traditional concessions. These include: joint ventures – equity joint ventures and contractual joint ventures (also described as "joint structures"), production-sharing agreements and a variety of service contracts (and contracts of work).

It is important to point out that in evaluating the relative merits of these forms, with a view to making a choice from among them, one must look behind the forms and labels. Thus it has been urged that:

> in looking at the evolution of concession policy and concession arrangements one should not be deceived by labels or forms. The modern day service contract, production sharing agreement or equity sharing contract may not be at any higher scale of evolution than straight concessions. It may simply be a new label. The host government is not necessarily going to make more profit from a service contract, a production sharing agreement, or an equity sharing agreement than it did under the old concession arrangement. The label is not the important thing here; it is the structure of the agreement and the fiscal arrangement within that agreement.[26]

This statement has, however, been rightly qualified by the observation that "labels may include notions of substance as they reflect a case of legal policy".[27]

As discussed in Chapters I and II, the development of new forms of agreement has been a reaction to the "traditional concession", which was perceived as an arrangement which worked overwhelmingly in favour of the companies. The main features of a traditional concession have been described thus:[28]

> It was a long-term agreement. 99 years was not untypical. It created enclaves and gave rise to exceptions to the laws of general application. Management "prerogatives" were left to the operating company. There was no government participation in the basic decisions, such as the rate of operation or marketing. The investor provided all capital typically, and was entitled to all profits, originally paying

the government a per ton royalty. Thus until the late 1950's the main interest that the government sought to secure was financial returns. The main thrust of government petroleum policy in the first decade after the Second World War was the increased financial returns. This was done mainly through introducing a form of taxation from which the government could appropriate 50 per cent of the profit. The 50/50 profit sharing formula thus marks the principal achievement of governments which sought to improve their relative position under the traditional concession agreement.

The traditional concession thus divested the state of its ownership over petroleum resources underlying the area covered by the concession. It is noteworthy that the transfer of ownership effected under a traditional concession meant that the exclusive right to explore and develop petroleum resources was granted to companies over large expanses of territory for substantial periods of time, and that the companies were thus invested with power to take critical decisions in all phases of operations. Thus, for all practical purposes, the companies could (as indicated in the earlier chapters) do as much or as little exploration as they wished. They retained exclusive rights over the concession area even if they chose not to undertake exploration activity. Thus, for a substantial period of time, the concession area passed beyond the control of the government and was not available for exploration or development by the government, either directly or through other companies. The companies had not only the right to determine the rate and extent of exploration, but in the event of a discovery being made, they retained the power to determine the rate and extent of development, and of the rate and level of production. They also determined price and controlled marketing. Thus the government only retained the right to receive a financial return originally in the form of royalties, and later in the form of royalties and taxes; all other rights stood transferred to the company. Consequently the government had little control over operations.

As has been noted in earlier chapters, the late fifties evidenced increasing concern by governments to secure control over petroleum operations. This concern was reflected in the assertion of the doctrine of permanent sovereignty over natural resources and led to the development of new types of legal arrangements under which oil companies could be involved without transferring to them ownership rights over the petroleum which may be discovered. Thus legal arrangements under which the government jointly owned the resources or retained ownership became an important objective in the new forms of agreements. The retention of ownership and the importance of these objectives was not merely "symbolic", as is sometimes suggested.[29] The non-transfer of ownership of the resources

110

to the company meant, in juristic terms, that the entire bundle of powers in relation to exploration, development and production of the resources, were retained by the government, *except* to the extent to which some of these were specifically conferred on the company. Thus, where a government was not initially divested of ownership, it was the companies which had to secure for themselves specific powers in relation to the different phases of petroleum operations.

It is true, however, that despite retaining ownership, a government may be divested of effective powers of control through transferring extensive powers of management to the company. In such a situation, the retention of ownership can truly be characterised as "symbolic". A comparative evaluation of agreements, therefore, must not only focus on provisions dealing with ownership, but involve an examination of the provisions which provide for management and actual control of operations. The government and the companies respectively are vitally interested in management, for while the government is interested in retaining as large a measure of control over operations as is possible, so as to be able to maximise benefits to the national economy from the development of its petroleum resources, the company seeks to have transferred to it as extensive as possible powers of management, so that it can enjoy the maximum freedom of action, and retain the freedom to take decisions in pursuit of its corporate objectives, defined within the framework of its global operations. The companies also attach substantial importance to retaining control over the actual disposition of the crude oil which is produced. Thus, it has been observed that companies are "apparently prepared to give up a number of percentage points in tax or royalty levels (or the price they pay to governments) for its share of crude oil in order to be certain that they will have the resulting crude oil to feed into their downstream operations".[30]

In evaluating the relative merits of different forms of agreements, therefore, an agreement must be seen as a package of mechanisms designed to secure certain interests and objectives. It should be appreciated that petroleum development involves a complex of objectives. There are certain basic general objectives, which each party to the agreement seeks to secure, while a number of distinct objectives are sought to be secured in each distinct phase of operations.

Among the general objectives which a government seeks to secure, the principal ones are maximisation of government take, control over operations, development of national capabilities in the field of petroleum operations, so as not only to be able effectively to monitor operations but also ultimately to undertake those operations directly, and

111

maximisation of benefits generally to the domestic economy.

No doubt, each type of arrangement involves a set of financial (and fiscal) provisions which determine how the operations are to be financed and how financial returns are to be shared between the respective parties in the event that a commercial production is attained. The next chapter will be devoted to a comparative evaluation of financial provisions. It must be said, however, that most existing evaluations tend to focus principally on a single aspect, namely financial provisions, so that the basis of comparison tends to be the financial return which accrues to the respective parties, that is, the government and the company, the returns to the latter being computed in terms of a DCF rate of return. These evaluations in focusing only on a single, albeit critical, feature, tend to obscure the importance of other features and the relative efficacy of mechanisms embodied in the different types of arrangement for securing other strategic objectives.

For purposes of comparative evaluation, it is useful to consider not only how effective the particular form of agreement is in terms of financial return to the parties, but to assess how effective it is in attaining certain objectives. Thus, it is important to keep in view both the general objectives to be secured and the specific objectives which are to be secured in each phase of petroleum operations and then to assess the effectiveness of the mechanisms incorporated under each type of arrangement to secure those objectives.

Exploration phase. The interest of the government in this phase is clearly to secure rapid and thorough exploration. This phase may, perhaps, be regarded as the most critical, since future development is premised upon the success of this phase. An area in which there has been a good deal of innovation is that involving legal mechanisms designed to ensure more rapid and thorough exploration.

While it is broadly true that "during the exploration stage there is less likelihood that contractors will pursue courses of action inconsistent with the interests of the host countries",[31] a number of situations have been pointed out in the previous chapter, where the interest of the government and the company may diverge, as, for example, where the geophysical data show indications of a small reservoir or one containing only gas. Further, a company faced with budgetary constraints in the context of its global operations may be inclined to commit less of its resources to a particular country than geological indications merit. Companies, therefore, aim to retain the maximum freedom to take decisions regarding the rate and extent of exploration.

As has been noted in the previous chapter, the relative bargaining strength of a government is at its weakest in the pre-exploration stage when the obligations of a company are being negotiated. The grant of a right to explore (or the right to explore and exploit) is what a government has to offer in exchange for the obligation undertaken by the company to carry out an exploration programme. It is, therefore, prudent for a government to require a company to submit its overall exploration programme and to negotiate minimum work obligations before exclusive exploration rights are granted to it, as is the case under the Norwegian or the British legal regimes.[32] Thus, before granting to a company an exclusive licence to explore (or to explore and exploit), a government should aim to secure from a company a commitment to a minimum work programme, spelt out in concrete terms. It may be difficult to elaborate a work programme in the pre-exploration stage, when the level of information available may be low. This does not, however, preclude the possibility of setting out certain minimum requirements with regard to the extent of geological and geophysical surveys to be carried out, the minimum number of exploratory wells to be drilled, and the minimum depths to which these should be drilled. Many of the individually negotiated agreements, which in overall terms appear to be advantageous to the government, are seen to have "weak" provisions stipulating the obligations of the company during the exploration phase.

Governments have an interest in resuming, within a relatively short time, control over areas where companies have carried out initial exploration operations and have not made any discovery, or have decided not to conduct further exploration: the more so, since there is evidence that subsequent exploration in areas abandoned by a company, has led to significant discoveries being made. The mechanisms by which this interest is secured are relinquishment or area-reduction provisions, which may be contained in the general legislation or in the agreement, or in both.

Since it is a basic government objective to maximise the rate and extent of exploration, governments are keen to retain power to determine the rate of exploration. It is also important for them to obtain data and information acquired by the companies in the exploratory phase. Governments, too, would like to have a say in decisions relating to exploratory drilling after the initial geological and geophysical surveys have been carried out, both as to the number of wells to be drilled and the depths to which these are to be drilled.

Thus, in comparing different regimes, a close scrutiny and analysis of the specific obligations of the company in the exploration phase is called for. The more effective provisions, which may be made to secure the interest of government in this phase are: a minimum work programme, specifying in some detail the nature of the exploration work to be carried out, the minimum number of wells to be drilled, with a stipulation as to the minimum drilling depths; minimum expenditure obligations (in addition to and not in lieu of incorporation of a minimum work programme), with an escalation clause, which provides for increasing the minimum expenditure obligation in the event of cost escalation (although this would not be necessary if a minimum work programme were stipulated and it were clearly stipulated that in the event of the cost of the work programme exceeding the minimum expenditure, the higher cost would be borne by the company); requirement to furnish to the government information acquired by the company in the course of exploration, including all geological and geophysical data and interpretations of such data; mandatory relinquishment, area-reduction and surrender provisions; sliding-scale area rentals; and fiscal incentives for accelerated exploration.[33]

Legal regimes may also be compared with regard to provisions made for governmental control over timing, direction, rate and level of exploration.

Development phase. It is important when comparing legal regimes to determine what conditions are required to be fulfilled before a company comes under an obligation to develop a discovery which has been made.

Following upon the initial exploration stage, a decision is to be taken about development. This is a critical stage where substantial investments have to be made in developing the reservoir. The cost of development is many times that involved in exploration. Under traditional concessions, the company was generally left with the discretion whether or not to develop a reservoir which had been discovered. As indicated in earlier chapters, problems arose where companies operating in the context of their global programme were unwilling to develop reservoirs, which governments were keen to have developed. Thus, under most of new arrangements, the company is required to submit all the relevant data to the government, and if that data indicates that a "commercial discovery" has been made, the company comes under an obligation to develop the reservoir. The basic issue, therefore, that needs to be resolved, is whether a discovery is to be regarded as commercially significant. Criteria for determining whether a discovery should be regarded as "commercial" or not vary from regime to regime. Some leave companies with greater discretion, while others introduce more objective criteria; others give the government power to require the company to develop if in the government's opinion the discovery is regarded as commercial. Problems arise when the government's view and that of the company about whether a discovery should be

114

developed diverge. This is likely in particular to take place with regard to marginal fields. A company having to choose between a number of discoveries globally may accord low priority to developing a field which may be regarded as marginal whereas for the government the development of such a field could provide significant economic benefits. Different legal regimes have sought to devise particular mechanisms to deal with the problem of marginal fields. Thus there are provisions which require the company to surrender any discovery which it is unwilling or unable to develop, leaving it then to the government to develop it through its own resources or in participation with another company. Thus, for example, upon refusal by an American independent company to develop a discovery made by it in Benin (Dahomey), the government prevailed upon the company to surrender the discovery which was developed with the assistance of a Norwegian company.[34] Other mechanisms include: sole risk clauses which entitle the government or the national oil company to undertake development of the discovery at its own risk; provision for development by the company under an agreement whereby the company will provide its technology and know-how, while the government would raise the necessary finance; sliding-scale royalties which provide for a lower royalty for smaller fields (or a progressive incremental royalty).[35]

Production phase. Upon development of a reservoir, the next stage is that of production. As has been pointed out in earlier chapters, the interest of governments and companies can diverge with regard to rates and levels of production, and with regard to adoption of certain production methods, involving such matters as secondary recovery procedures, flaring of associated gas and anti-pollution measures.

It is important therefore, to analyse the provisions made under each regime to determine the extent to which the government can influence or affect decisions relating to such critical matters as rates of extraction, levels of production, adoption of secondary recovery procedures, flaring of gas, etc. It is as important to determine the extent to which standards are laid down to regulate petroleum operations. In this context, a question that needs consideration is the adequacy of a simple general standard such as "good oilfield practice". The problem has been well stated, thus:[36]

> The two operating companies . . . are bound by the terms of the agreement to "conform with good oil industry practice . . . in conserving the deposit" . . . in the case of production, such standards are lacking everywhere in the world, except in the oil-bearing states of the United States. The reason is obvious. Only in these states have the problems of production and conservation been sharply contested and settled . . . We know that [in oil and gas leases in the US] there is an implied

covenant to protect against waste and drainage. Does this apply in the case of Iran? . . . If it does not what law would control? In the United States the law is not uniform among the oil-bearing states.

The limits of "good oilfield practice", as a standard for regulation of production, or the rate of depletion, are brought out in the answers of the representatives of the British Department of Energy to questions put to them by members of a Parliamentary Select Committee, thus: [37]

Q. Will you confirm again that at this moment the British Government would have no power at all to regulate the rate of extraction in accordance with what the Government felt are our national economic needs?

A. *The only power that the Government has under the existing licences arises from the obligation to conform to good oil-field practice, and I think it is fair to say that alone does not give the Government a very strong or positive power to alter the rate of extraction materially . . .*

Q. You mentioned good oil-field practice. Is a list of these practices available in any written form or is it part of the folklore of the exploration industry?

A. There are publications by the Institute of Petroleum which give a fair indication of what is normal good practice in that industry . . . It is an industry document to regulate the technical affairs of the industry . . . to establish common practice between companies . . .

The fact that the licences impose a condition of conforming to good oil-field practice does mean that the Government has a say in this. I do not think we have ever had occasion to use this, but the way we go about it is that our technical people do have discussions with each of the licencees or each of the operators about their plan for developing a field, and if it seemed to our technical advisers that the operators were going about this in an irresponsible or technically defective way, I am sure we would have something to say about it. *What I think is much more doubtful under the existing powers is whether we could say for conservation reasons that such-and-such a field should be developed very much more slowly than the operators plan. I think it would be questionable whether such a point would fall within the licensing clause about good oil-field practice.* [italic wording added]

Thus, it is evident that little reliance can be placed on "good oil-field practice" to derive objective criteria by reference to which a government can call upon a company to reduce depletion rates, or insist upon adoption by it of particular production methods, say, involving secondary recovery procedures, re-injection of gas into reservoirs, or flaring of associated gas. Company representatives take the view that since these matters have financial and economic implications, they are not the kind of technical questions which can be answered by reference to the standard of "good

116

oil-field practice".[38] Further, such practices as are recognised have developed in the background of the US petroleum industry, and have been formulated by the American Petroleum Institute, which represents the industry's interests. Government representatives have, therefore, expressed the view that these standards provide inadequate safeguards, so that governments should provide for more extensive regulations and frame their own detailed regulations if they are to effectively regulate petroleum operations.[40]

Marketing. The disposition of the crude oil produced is a matter which raises issues of some importance both to the government and the companies. Companies have always attached importance to control over crude oil, for it has been observed:[40]

> Although an economist would be tempted to say that control over the product is secondary to the question of who pays how much to whom, that fundamental economic insight leaves out of account the great importance integrated oil companies have placed on obtaining a secure source of crude oil to keep their downstream refineries and distribution systems operating at or near capacity. This aspect of the behaviour of large international oil companies raises fascinating questions of industrial organization, but for the present purposes it will suffice to observe that these companies act as if control over crude is highly important.

Governments have an interest in ensuring that the best price is obtained on export sales. In addition, some governments have an interest in appropriating out of the production such quantities as are needed for domestic consumption, and are inclined to seek these supplies at a special formula price.

Control over marketing acquires importance also in relation to pricing. In nearly all cases, price is relevant for the computation of the government take. Under traditional concessions and other arrangements where the company was free to market the oil, problems arose with regard to the sale of oil to its affiliates at "transfer prices", which governments regarded to be artifically deflated. This led governments to insist that regardless of the realised price, a notional price should be fixed, as the basis for realisation of taxes or other dues payable to the government. Thus, arose concepts such as "posted price" in the Middle East, "the tax reference price" in Venezuela, and more recently the "norm price" in Norway.

A different mechanism found in some of the new agreements, gives the government the option to market the oil, if it can obtain a better price than that which can be secured by the company. A more fundamental change is reflected in the service contracts under which the right to market oil is vested in the government, and the company has only the right to

117

buy a certain percentage of oil at a formula price. These mechanisms merit examination in a comparative study of petroleum agreements.

A comparative study of petroleum development agreements must, therefore, seek to look behind the labels and differences in form and examine how a number of specific matters are dealt with. Thus in looking at these arrangements the following questions may be asked:

Ownership and management

1. How is the question of ownership of petroleum dealt with?
2. Is the retention of ownership by the state/government symbolic? In other words, does the government, while retaining ownership, transfer substantial powers of management (powers to take critical decisions) to the oil company? Where is the balance struck between supervision and control by the government to secure national policy objectives and the company's freedom to operate so as to secure its corporate objectives?

Exploration phase

1. To what extent is the company obliged to furnish to the government information and geological and geophysical data gathered by it during exploration?
2. To what extent is the company required progressively to relinquish or reduce the area held by it for purposes of exploration?
3. Is there a minimum work obligation imposed on the company? To what extent does the government retain the power to control or supervise the extent and rate of exploration?
4. Is there a minimum expenditure obligation imposed on the company?
5. Is the company required to furnish a performance bond to guarantee performance by it of its obligation?
6. Are there any fiscal or financial provisions, which provide an incentive for rapid and thorough exploration?

Development phase

1. What provisions are made for development of any reservoirs which are discovered?
2. What conditions are to be fulfilled in order for a company to come under an obligation to develop such a reservoir? How is a "commercial discovery" defined?
3. What is the extent of the government's control over the development decisions of the company?

118

Production phase

1. How much control does the government have over decisions and policies relating to production?

2. To what extent can it influence or determine decisions relating to the rate of extraction, levels of production, adoption of secondary recovery procedures, flaring of gas, etc.?

3. What standards are laid down to regulate petroleum operations by the company and how far do these enable the government to regulate and effectively supervise the operations of the companies?

4. To what extent does an obligation to comply with "good oil field practice" provide the basis for effective supervision by the government of petroleum production operations? Is there provision for more extensive regulation of production operations?

Marketing

1. To what extent does the government influence/determine marketing decisions, including pricing?

2. What provisions are made in relation to sales by the company to its own affiliates?

Financial provisions

1. What is the nature of the financial provisions? In what forms does the government receive its financial returns under the arrangement in question?

2. Does it receive a bonus at the time of signature? Is the magnitude of the signature bonus so large as to affect the availability of funds for exploration?

3. Is there provision for payment of production bonuses?

4. Are there provisions for payment of a royalty and taxes? Is there a provision for the payment of "rentals"?

5. Is the amount of tax (royalty or other payment) payable by the company liable to be varied from time to time by legislation?

Other benefits to the national economy, including transfer of technology and skills

1. What provisions are there for transfer of technical and managerial expertise?

2. What provisions are made for training and for participation of nationals in different spheres of petroleum operations?

3. What provisions are made for employment of nationals, preference in the purchase of goods and services from nationals, supply of petroleum for the domestic market, on establishment of refineries and other industrial installations, etc?

4. What provisions are made for other benefits to accrue to the economy?

The new forms of petroleum agreements, having emerged in response to assertions by states of sovereignty over their natural resources, provide for varying degrees of "state participation". Thus, under each of them ownership of the petroleum resources in the ground remains with the state, and at best provision is made for "sharing" of the production at the well-head between the government and the company. Greater powers of management and control over operations are vested in the government. The duration of the agreement is shorter and the size of the areas covered by the agreement considerably smaller than under the traditional concessions. Relinquishment and area reduction provisions are contained in each of them so that the area initially granted under the agreement is progressively reduced as exploration proceeds, so that, unless a commercial discovery is made, the entire area is resumed by the state within a period rarely exceeding 10–12 years. Most agreements contain minimum work and expenditure obligations to be discharged by the company during the period of exploration, and impose obligations on the company to train and employ nationals, to procure goods and services locally, and to undertake to supply crude oil required for domestic consumption, often at a "cost-plus" formula price.

It will be useful to examine the new forms of agreement — joint ventures, production-sharing agreements, service contracts (and contracts of work) — in the framework outlined above so as to assess the extent to which changes in form represent changes in the substance of the relationship between governments and companies, and to see how in fact these agreements have worked in practice.

JOINT VENTURES

Governments as they became conscious of the need to have a say in the actual management of petroleum operations and in the taking of critical decisions pressed for "participation". The oil companies preferred partnership to be limited to "profit sharing", while governments pressed for greater control over operations.[41] Since the nearly total control exercised by companies was based on its ownership of petroleum under traditional concessions, the urge to share in the control of operations was reflected in the desire to share the ownership.

120

Equity joint ventures

The joint venture agreement pioneered as a new form of petroleum development arrangement by ENI, the Italian State Oil Corporation, was a response to the need felt by governments to share in the ownership and control over operations. The essence of an equity joint venture is that the host government (or its national oil company) and the foreign oil company form a partnership whereby they establish an operating company in which each own 50 per cent of the shares.

The earliest joint ventures were those established by ENI and certain Egyptian concerns and by Agip Mineraria, an ENI subsidiary, and NIOC of Iran, both in 1957.[42] These were equity joint ventures, under which a joint stock company was established, in which each partner owned 50 per cent of the equity. The company was charged with the functions of exploration, production and sale of any crude oil or other hydrocarbon that was produced. It has been observed that:

> the symbolism of the joint venture lies not simply in the jointness and in the existence of an entity called a state oil company, but also in the fact that territory is not licensed solely to a foreign oil company.[43]

It is noteworthy, however, that although ownership of any petroleum discovered was joint and the operating company was jointly owned, the entire risk capital for exploration was to be furnished by the foreign partner. In the event that no commercial discovery was made, the loss was to be *exclusively* borne by the foreign partner. In the event of a commercial discovery being made, the jointly-owned operating company would be reimbursed out of the revenue earned.

The joint venture system was projected as one which made the host country a real partner by involving it in the management of petroleum operations. This feature of the joint venture was thus described by ENI:[44]

> The association (joint venture) system is completely different from the traditional concession system. When the fields discovered are being developed, the host country takes a direct part in running the joint enterprise through its own managerial, administrative and technical staff and this ensures that the country's interests are represented in all decisions affecting the formation of the oil revenue, while at the same time the staff are acquiring training and experience.

The formal structure of management was based on the parity of shareholding of NIOC and AGIP, which was reflected in the composition of the Board, half of the directors being appointed by NIOC and half by AGIP

121

(Article 4). In practice, however, the effective technical management is vested in the foreign company. Article 4 of the NIOC/AGIP agreement, while providing for equal rights with respect of the formation of Board of Directors and Board of Auditors, stipulated that the Managing Director would be appointed by AGIP. It was further provided that the technical management of the company would be entrusted to personnel designated by AGIP and/or appointed by the Managing Director.

With regard to the exploration phase, a "weak" provision was made. It was provided (Article 8) that AGIP would be in charge of formulating exploration programmes as well as determining conditions of execution. It was AGIP which would prepare and draw up exploration programmes "after consulting with NIOC" (article 8). AGIP would execute the work programme under its own supervision, but information would be furnished to NIOC, as AGIP was required to submit detailed reports on the progress of the work to the jointly owned company and to submit a final report. The minimum work obligation (Article 18) required that geological and geophysical exploration should start within 6 months from the date of the agreement and that drilling operations should begin whenever it was deemed appropriate, and in any case before the expiration of 4 years from the date that the agreement became effective. A minimum expenditure obligation was also stipulated, namely 22 million dollars over a period of 12 years (Article 19).

A relinquishment clause (Article 26) provided that at the end of 5 years from the commencement of exploration the area covered by the agreement would be reduced by 25 per cent. At the end of the ninth year, another reduction of 25 per cent was to be effected, with AGIP having the freedom to select the portions it chose to retain. At the end of the twelfth year, AGIP could retain only those areas in which commercial quantities of oil had been discovered.

The definition of commercial quantities embodied in the ENI-AGIP agreement was as follows:[45]

> The yield capacity of a petroleum field in a commercial quantity will under prevailing conditions be estimated when the amount of oil extraction reasonably foreseeable is such that when the cost price of delivery to seaboard, calculated on the basis of production costs plus transport and handling charges and an additional 12½% of the posted price payable as a minimum for tax and duties to the Iranian government is deducted from the posted prices, of a similar kind of petroleum, would leave a reasonable margin of profit.

This agreement did not, however, incorporate a mechanism which would enable one party to proceed to development at sole risk, in the event that there was a difference between the parties on the assessment of the commercial prospects of a discovery. It is noteworthy that such a problem in fact arose under the AGIP-NIOC agreement when in 1960 AGIP struck oil off shore in two wells with a daily yield of 3500 barrels each. It is reported that "it was not until May 1962, and this only after considerable efforts on the part of ENI, that NIOC acknowledged that the field was commercially exploitable", the reason being that[46]

> NIOC was hesitant to recognise the commercial exploitability of the find, because it would have then been obliged to put up half of all the subsequent development expenses, as well as to reimburse AGIP for half its exploration expenses under the joint-venture agreement.

This is a problem which inevitably arises when a party which is to make substantial investments in development is faced with a discovery about the commercial prospects of which it takes a different view from the other party involved. In this case it was NIOC which was reluctant to undertake the investment in what from its point of view did not appear to be a commercially attractive venture. More often it is the government or the national company which is keen to develop a discovery, which a company does not find commercially attractive.

A very formal mechanism was incorporated in the AGIP-NIOC agreement (under which SIRIP was the joint operating company) to deal with situations of deadlock, thus:[47]

> Each of the two parties to the joint venture is required to deposit one percent of the shares of the operating company in the Union des Banques Suisses in Zurich. The deposit of two percent of the capital stock of SIRIP is made jointly under the joint signature of the two principals, and the designated Swiss bank is given a joint and irrevocable mandate under the terms of the agreement, and for a period equal to its duration. If this bank refuses or relinquishes the mandate, it may nominate, to succeed itself, another bank of equal status.
>
> A person is nominated in advance for each year before 31 December of the preceding year, by mutual agreement of the two parties, to exercise the voting power of the shares held in mandate. Failing such agreement, the President of the Cantonal Tribunal of Geneva shall nominate such a person from among nationals of a third country, other than the countries of the parties to the joint venture, provided the nominee does not have any interest in any petroleum undertaking and is not a government official of any sort.
>
> The mandatory bank undertakes to deliver the admission card for each general meeting, in respect of the deposited shares, to the person so nominated or chosen. This person then exericses his voting mandate in the general meeting. In doing so, he is under obligation to act with strict regard to the interests of the joint venture within the provisions and spirit of the agreement.

123

This mechanism was not satisfactory, for while it provided a formal procedure for resolving a deadlock, it was more akin to an adjudicatory procedure than one for arriving at a consensus in the interest of the joint venture. This mechanism did not commend itself to the parties in subsequent joint venture agreements.

A more efficient mechanism was developed and incorporated in later joint venture agreements to deal with situations where the parties did not agree on the commercial prospects of a discovery. A clause was included which enabled either party to undertake to develop a discovery at its "sole risk". Thus, the "sole risk" clause incorporated in the Pan American-UAR Agreement of 1963, under which operations were entrusted to the joint operating company, the Fayoum Petroleum Company (FAPCO), provided the following framework for dealing with this situation:[48]

In the event of the failure of the board of directors of the operating company to approve any drilling, exploration or other development operations under the agreement, due to a difference of opinion between the two principals, the party supporting the specific operation may go ahead on its own after following a specific procedure. Such party, referred to by the agreement as the "proposing party", should propose in writing to the other party, referred to as the "non-proposing party", that such particular work be done; and a copy of the proposal should be furnished to FAPCO. The non-proposing party may, within 60 days from the receipt of such proposal, elect to participate equally in the cost of the said work. If it does not do so, then the proposing party shall notify FAPCO of the result, such notice serving as an authorization by the said party to the operating company to carry out the particular operation at its sole risk and responsibility. In such operations, FAPCO will be acting as the sole agent of the proposing party, and not of the joint venture . . .

The proposing party is then entitled to receive and own all the petroleum extracted in the sole-risk operation it had financed unilaterally. It continues to do so until it has realized the following revenues:
1. For each wildcat well an amount equal to:
   (a) The costs and expenses incurred and paid by the proposing party for the drilling, testing, equipping and completing of the well, plus
   (b) three hundred per cent of such costs and expenses.
2. For every other well, an amount equal to:
   (a) The costs and expenses incurred by the proposing party for the drilling and development operation, plus
   (b) one hundred percent of such costs and expenses.
3. For other projects, an amount equal to:
   (a) The costs and expenses incurred by the proposing party, plus
   (b) seventy-five percent of such costs and expenses.

124

After the proposing party recovers these amounts, the operating company serves notice on the non-proposing party of such recovery. In this event, the non-proposing party is entitled to participate in the sole-risk operation with the proposing party upon making a payment to the latter party of 50 per cent of the total expenses incurred. If such payment is made within 90 days of the receipt of notice, then the well, wells or other projects involved shall pass to the joint venture, and be jointly owned and controlled by Pan American and EGPC under the agreement.

If the non-proposing party declines to pay its share of the said expenses, then the proposing party acquires permanently the ownership of the project or projects so developed at its sole risk, with all their installations and facilities. Thereafter, FAPCO continues to operate and maintain such projects as the agent of the proposing party, at its sole cost and expense. All the petroleum produced from such wells is owned by the proposing party alone.

On the other hand, if any well drilled at the sole risk of the proposing party turns out to be a dry hole, then it is sealed and abandoned by FAPCO at the sole cost and expense of this party.

An instance where the sole risk clause was invoked was the Abu Qir gas discovery in Egypt by Phillips in 1969, which the company did not regard as commercial, but which the government decided to develop at its sole risk in order to meet domestic requirements.

The "sole risk" clause has been commended as "one of the admirable instances of flexibility made possible by the joint venture structure".[48]

Another instance of flexibility is provided by the mechanism, embodied in the Pan American-UAR Agreement of 1963, to deal with a situation where a project is approved by both parties but one party is either unable or unwilling to share in financing it. This mechanism operates in the following manner:[49]

After a project is approved by the board of directors of the operating company, the only party willing to pay, called "the paying party", may advance to the operating company the necessary funds. After the completion of the project in question, the other party is entitled to use the facilities of the project equally with the paying party, and starts sharing with it equally all the expenses of operation and maintenance following completion. However, in order to compensate for the expense incurred by the paying party alone, this party has the right to recieve, from the non-paying party, an amount equal to half the costs and expenses not shared by the non-paying party, plus 75 percent of this amount. These refunds are paid monthly at a rate equal to double the amortization rate fixed by the joint-venture agreement.

However, the non-paying party may avoid the above provision if it is able to make payment on the approved project within nine

125

months from the date it was due. In such event, the paying party is entitled to an interest of one-half percent per month on such payments to compensate for the delay.

After the above payments are made by the non-paying party, this party acquires an interest in the approved project or investment equal to that of the paying party, and the project is operated by the operating company as a part of the joint venture.

With regard to the production phase, an obligation was imposed on the joint company to use all its possible efforts in order to raise to a maximum the sale level of petroleum and for this purpose to develop the production of such fields so that production was achieved within the limits compatible with the most modern technical procedures in the oil industry (Article 12).

A particular problem presented by the joint venture is the relation between production and "off-take" by the respective parties. So long as both parties lift oil in proportion to their equity interest, the arrangement presents no difficulty. It is, however, pointed out that:[50]

> The problem arises when there is a persistent underlift. The obvious solution would be for the over-lifter to provide the necessary additional capital, but this would create the problem of altering the equity share of the two sides. Consequently, the overlifter must be able to buy crude from the persistent underlifter at a specific price.

This problem is dealt with by different mechanisms embodied in the provisions relating to marketing. Under the AGIP-NIOC agreement, the operating company was responsible for formulating sales policy and marketing the oil produced. It was to market the oil in the interest of the joint venture and was in principle empowered to sell to any party, and was not legally obliged to ensure supply to either party. The Pan American-UAR Agreement of 1963 introduced a provision whereby the company undertook to help in marketing the portion of the government's share which the latter could not dispose of, in the following terms:[51]

> in the event EGPC is unable to take in hand or separately dispose of all or a reasonable share of its entitlement of liquid hydrocarbons within a reasonable period following the preparation and furnishing of FAPCO's first forecast . . . EGPC and Pan-American shall meet to discuss equitable arrangements under which EGPC could nominate and take larger quantities of its entitlement. Without incurring any obligation, however, Pan American shall consider means by which it could assist EGPC in achieving such larger nominations on a basis mutually acceptable to EGPC and Pan American.

Subsequent joint ventures, in particular the contractual joint ventures considered below, incorporated provisions under which a party could buy at a special price the oil which the under-lifting party had failed to lift.

126

Under the Kuwait-Hispanoil Agreement of 1967,[52] which contemplated the establishment of an equity joint venture, Hispanoil was not only committed to a substantial off-take but also to sell up to a maximum of 20,000 barrels a day from the Kuwaiti share of the production.

The financial provisions under the AGIP-NIOC agreement required that reimbursement of prior exploratory expenses be made out of oil revenue from the field in question and, as indicated above, only 50 per cent of the total costs was to be paid to AGIP, the other 50 per cent being paid to NIOC, though NIOC had not contributed towards the exploration expenses. It had been rightly observed that AGIP was in effect paying NIOC a rental equal to half of all exploration expenses whenever a commercial discovery was made.[53] The joint company was to pay the Iranian Government a 12½ per cent royalty and 50 per cent of net profits after deduction of the royalty. AGIP paid a further 50 per cent tax on its 50 per cent interest in the joint company. Thus AGIP paid taxes at the rate of 75 per cent after the deduction of a royalty of 12½ per cent on its share of oil production.[54]

Contractual joint venture (joint structure)

A form of joint venture extensively adopted is characterised as a "contractual joint venture" or a "joint structure". This was the type of structure established, for example, by the agreement between the National Iranian Oil Company and Pan American Oil Company of 24 April 1958, or the agreement of 17 January 1965, between NIOC, AGIP, Philips and the Oil and Natural Gas Commission of India.[55] Under a contractual joint venture (or joint structure) the partnership is not constituted into a joint stock company, and thus does not assume a separate corporate identity. The relations between the parties are governed by the terms of the partnership contract. The petroleum produced is not jointly owned. Each partner owns a 50 per cent undivided share, and thus directly owns its own share of the production; thus, Article 23 of the 1965 agreement provides:[56]

> Petroleum produced from the area shall be owned at the well-head 50 per cent by the first party and 50 per cent by the second party.

Thus the 1965 agreement by providing for "the direct ownership by the government of part of the production introduces a subtle but significant change in relations between the government and foreign companies".[57] Under the AGIP/NIOC equity joint venture of 1957, it was the operating company which was vested with ownership of the oil and thus with the power to sell the oil to NIOC and AGIP on conditions acceptable to it.

It was free to sell to other purchasers at conditions no less favourable to itself; consequently, the foreign company was not legally assured of security of supply. Under the joint structure arrangement, each partner directly owns and is, therefore, free to dispose of its share of the oil.

An added advantage of this type of arrangement for an American company is that if it can prove to fiscal authorities that it has direct ownership of its part of the production, it gains a number of significant fiscal advantages, such as the benefit of deduction of intangible expenses and a depletion allowance.[58] Further, a joint structure not being a joint stock company, is not subject to the restrictions imposed by the domestic company law of the country in which operations are being conducted. It has thus been observed that:[59]

> In the case of a joint structure the ties with the local legal system are looser, almost any rule may be adopted in the contract at the parties' will. Contractual joint ventures are usually a more flexible form of cooperation than equity joint ventures.

Under a joint structure agreement, management is entrusted to a non-profit making joint stock company, which has no balance sheet and which is not subject to taxation. In some cases, the joint operating company is established at the outset. During the exploratory period, it acts as the agent only of the foreign partners, while in the later phases of development and production, it acts as the agent of both parties.

A possible handicap of a non-profit-making operating company is that it may not have at its disposal any reserves and could face operational difficulties if it exceeded the precise budgetary limits approved by the partners. In practice, however, the budget can be planned so as to leave a sufficient margin of financial freedom to the management. Also as agent, it has the power to take any action necessary in an emergency; it has been observed that:

> The legal structure established by the 1965 agreement is not expected to hamper efficiency of management, although it is not denied that in certain exceptional circumstances it could be an obstacle.[60]

There is parity of representation on the Board of Directors, 50 per cent of the directors representing NIOC and the other 50 per cent the foreign partner. The Chairman is to be appointed from among the directors appointed by NIOC, and the Vice-Chairman from among those appointed by the foreign partner. The Managing Director, the Chief Executive, is to be nominated by the foreign partner. The day-to-day conduct of operations and technical management is in effect left to the foreign partner.

The annual budget and production programme for each year has to be approved by the Board. Provision is made for constituting an Audit Board consisting of one auditor nominated by each party.

The provisions regarding the exploration phase embodied in joint structure agreements have been progressively strengthened over the years. The 1958 NIOC-Pan American agreement provided for relinquishment over a 12-year period, so that at the end of 12 years from the date of signing of the agreement, the joint venture could only retain areas in which commercially exploitable fields had been discovered (Article 3). The 1965 joint structure agreement,[61] in Iran, provided for relinquishment of 25 per cent of the area not later than the end of the fifth year from the effective date (date of signature of the agreement), and of 50 per cent of the area not later than the tenth year, to be followed by relinquishment of the remaining area, less the area covered by any commercially exploitable discovery. In the 1971 agreements,[62] the relinquishment programme was substantially accelerated and the total exploration phase was reduced to 6 years. Thus, 25 per cent of the total area was to be relinquished within 3 years, another 25 per cent before the end of the fifth year, and the remaining area by the end of the sixth, except for any area in which a commercially exploitable discovery had been made.

The joint structure agreements provide for a signature bonus, and for minimum expenditure obligations during the exploration phase. The 1965 agreements spelt out minimum work obligations which required the foreign company "to exert its utmost efforts to explore the area to the maximum extent consistent with good petroleum industry practice" and to commence drilling at least one well within a maximum period of 18 months from the effective date. The 1971 agreements contained minimum work obligations, together with the requirement to submit to NIOC progress reports on exploration work done and a comprehensive final report.

The exploration risk in all these agreements was borne by the foreign partner, who was responsible for bearing the entire cost of exploration, which was not reimbursible unless a commercial discovery was made. In the event of a commercial discovery, the foreign partner is refunded 50 per cent of all its exploration costs and expenses in annual instalments, as at a certain rate per exported barrel.

Under the 1958 agreement, "a discovered petroleum reserve" was regarded as "commercially exploitable":[63]

> if the quantity of petroleum reasonably foreseen as derivable therefrom is such that delivery of petroleum at the seaboard shall be possible on the following basis: if the cost of production is increased

by transport and loading charges as well as a sum equal to 12½% of the applicable posted price for petroleum of similar quantity, and if the resultant figure is deducted from the said posted price, there should be left a net profit of not less than twenty five per cent of the applicable posted price.

Thus, "the minimum profit serves the purpose of guaranteeing the royalty due to the government".[64]

As pointed out above, these joint structure agreements, unlike the NIOC-AGIP equity joint venture, contained a "sole risk" and a "unilateral financing" clause. There is also a provision in the 1965 NIOC agreement that if and when NIOC requires, the foreign partners will, in addition to their own share, also provide NIOC's 50 per cent share of all the expenditure required for development and exploitation of commercial fields until the date of commercial production. NIOC could, in that event, pay to the foreign partners interest equal to the rate of discount of the US Federal Reserve Bank plus 1.5 per cent. The foreign partner would be repaid the amount in sixteen equal semi-annual instalments, as from 6 months after the date of commencement of commercial production. In the production phase, the parties jointly assume the responsibility for all expenditure required for operations.

The foreign partner in effect controls the technical aspect of production operations, though with the increasing technical capabilities of the national company, and an express provision investing NIOC with powers of supervision, the freedom of decision-making of the foreign partner has been progressively restricted. The 1971 Iranian joint structure agreements require the operating company "to comply with principles of conservation of natural resources" and "in the conduct of operations" always to be "mindful of the best interest of Iran"; it is expressly provided that "NIOC shall exercise all necessary controls for supervision" required to ensure full compliance with those principles.

The marketing arrangements under this type of joint structure provided specific safeguards for each party. Thus, each party was entitled to take from the operating company half of the quantity available for export. In the event that NIOC was not able to find profitable markets abroad, the foreign partner undertook to export and sell NIOC's 50 per cent share of crude at a formula price, which in the 1965 agreement was a "half-way" price, equal to the total cost of the barrel plus half the difference between the posted price and the cost.[65]

The financial provisions in the joint structure agreements, while proceeding on the basis of equal sharing of profits between partners, as under equity joint ventures, in effect yielded a result more favourable to the government. Under the NIOC-AGIP equity joint venture, the net profits of the joint venture were paid one half to Iran and the other half to AGIP and NIOC; thus, the split was 75/25 in favour of the government. Under the 1965 joint structure agreement, the joint structure relationship did not entail any tax obligation; NIOC and each of the foreign companies were separately subjected to taxation and had to draw up separate balance sheets. This mode of calculation brought about "a split that is even more favourable to the government, that is about 80–85 per cent for the government (since) gross receipts . . . are not calculated on the basis of the posted price . . . which is about 30 per cent higher [than the market price]".[66]

It is noteworthy that the 1965 agreement subjects each of the parties to taxation in accordance with the Iranian income-tax laws "as they may prevail from time to time", so that the government reserves the right to change the tax rates with a proviso incorporated in the agreement, whereby the government guarantees to the foreign company that it would not be "subject to rates of income tax or other provisions governing net income which are less favourable than those to which other companies, engaged in similar operations in Iran, which together produce, or cause to be produced, more than 50 per cent of Iranian crude oil, are subject . . ." (Article 30).

Joint ventures with special features

A number of joint ventures contain special features which merit attention — (a) the Saudi-Arabian joint ventures; (b) the Libya-Auxirap Agreement of 30 April 1968; and (c) the Carried Interest Joint Ventures (with special reference to those developed in the Netherlands and Norway).

(a) *The Saudi-Arabian joint venture agreements*. Under the Saudi-Arabian-Auxirap Agreement of 4 April 1965, the foreign partner was to carry out exploration at its risk. In the event of a commercial discovery, Saudi Arabia had the option of 40 per cent equity participation, and a joint operating company was to be established. While Saudi Arabia was to have a 40 per cent interest in the venture, there would be parity of representation in management. The joint venture would embrace all phases of operations, including those "downstream" — refining, transport and marketing.

The Saudi-Arabian AGIP Agreement of 21 December 1967,[67] is based on an initial grant of a 3-year Exploration and Prospecting Licence from the Saudi-Arabian Government to its national company, Petromin. Under Petromin's agreement with AGIP, the former assigned its exploration licence to AGIP, which was to have "exclusive management and control of all operations" under the Exploration Licence (Article 3).

An Exploitation "concession" for a period of 30 years would be granted if an application were made within 90 days from the date of discovery of oil in commercial quantities (Article 8 of the Exploration Licence), which was defined thus:[68]

> ... the date of discovery of oil in commercial quantities shall be the date of testing a well within the Licence Area capable of producing by itself or together with wells already drilled and tested within the same area, in accordance with first-class oil-field practice, a minimum of twenty-five thousand (25,000) barrels of 42 US gallons at 60 Fahrenheit) per day for a period of thirty (30) consecutive days.

In the event of such a discovery, Petromin would have the option "to retain thirty per cent (30%) of the undivided economic and beneficial interest therein" (Article 3) and if it exercised the option, it would pay to AGIP 30 per cent of the initial expenses (which would include all exploration costs but exclude bonus of rental payments made to the government). A "sliding-scale" participation formula (based on levels of production attained) is built into the agreement (Article 9) so that if crude oil production from the concession area averaged 300,000 barrels a day for a period of 90 consecutive days, Petromin would be entitled to acquire a further 10 per cent economic and beneficial interest, on payment of 10 per cent of the initial expenses and of development and intangible drilling expenses (after making allowances for depreciation). Having exercised this option, Petromin would have the right to acquire a further 10 per cent interest in the event that crude oil production averaged 600,000 barrels per day for a 90-consecutive-day period.

The conduct of operations was to be entrusted to a non-profit making Saudi-Arabian company, each party subscribing to the equity of that company in proportion to the interest held by it. The operations would be conducted in accordance with agreed "rules and procedures".

AGIP undertook the obligation to market Petromin's share of crude oil production, if required to do so. It was agreed that at least 35 per cent of the total production would be sold to independent third parties, and that payment to Petromin for sale of its share of crude oil would be calculated on the basis of the weighted average realised price in sales to third parties. If the realised price fell below the "expected third party price", which would be determined in advance by agreement, Petromin could require that production be reduced to a figure not less than "the economic level", which is defined in terms of a level of production which would assure AGIP of a return after income tax of 10 per cent per annum on cumulative net worth.

132

It is noteworthy that when a further area was included in the concession area by an addendum dated 10 April 1968, the obligation of the foreign partners during the exploration period were specifically spelt out both in terms of minimum expenditure and minimum work obligations (including drilling of wells to certain depths within the stipulated period).

(b) *The Libya-Auxirap Agreement of 30 April 1968.*[69] Under this "joint-structure" agreement, exploration and appraisal work was to be carried out by the foreign partner, under the general supervision of a management committee of six persons, of whom initially two would represent the government and four the foreign partner. A "sliding-scale" participation formula provided that if production were to exceed 200,000 barrels per day, the government's share would increase by 1 per cent for each 14,000 barrels per day, until a production level of over 500,000 barrels a day was reached, at which stage the percentage interest of each party would remain fixed at 50 per cent.

The obligations of the foreign partners in the exploration phase are elaborately spelt out. Minimum expenditure and minimum work obligations are set out as is the requirement for progressive relinquishment of the area covered by the contract.

A field was to be regarded as "commercial" so as to give rise to the obligation to develop it when it could be capable of producing 90,000 barrels per day. If it could not, then the joint venture as such would not undertake its development, unless the foreign partner agreed. If the foreign partner did not agree to do so within 18 months, then the national company could negotiate with the foreign company a new basis for development of the discovery, or alternatively proceed to develop it on its own.

The parties were entitled to lift in kind quantities up to the amount of their share of the "technical maximum production", which was defined as "the maximum quantities of petroleum that could be economically produced, according to good petroleum field practice, from the production wells of the developed field".

The foreign partner undertook to market the government's share of oil at weighted average prices in consideration of payment of its "marketing expenses" at a flat rate of 2 per cent of the sales proceeds.

A noteworthy feature of this agreement is also the elaboration of accounting procedures and methods of calculation relevant to the computation of the tax payable to the government. Thus for example it expressly specifies items which shall not be treated as deductible expenses; these include foreign taxes paid on Libyan income and interest or other charges incurred in connection with financing operations or preparations for operations in Libya.

(c) *The carried interest system.* In any joint venture, which provides that the company will finance the exploration and bear the risk, so that in the event of no discovery being made, the loss will be borne exclusively by it, but at the same time provides that the government can acquire an equity interest if a commercial discovery is made, the company can be said to "carry" the government's "interest" during the exploration phase. In the earlier joint ventures, once a discovery was made, the government or the national company was obliged to "farm in", that is, buy its share in the venture, undertaking to reimburse the company for its proportionate share of expenses, and to contribute towards the financing of the joint venture in proportion to its share in it. A further development was represented by the Saudi-Arabian and Libyan agreements which gave the government or the national company an *option* to "farm in", if it elected to do so. They also introduced a sliding-scale participation formula so that the larger the discovery, the higher the share that the government or national company could acquire.

The "carried interest" formula was introduced into the North Sea region by the Netherlands. By the Decree of 27 January 1967, the government was given an option to acquire up to a 40 per cent interest in natural gas discoveries but the government's declared policy is that it will participate only when the find is substantial;[70] more recent legislation contemplates a similar option to be given to the government in respect of commercial oil discoveries.[71]

In Norway, the discovery of condensate in the Cod field in 1968 encouraged the government to introduce provisions for participation in the licences granted in 1969. Under the 1969 licences provision was made for two types of participation: net profit sharing and carried interest. The former provided for a share of net profits to accrue to the government or the national company. This was akin to an income tax, with the difference that when the state assigned its interest to the national oil company, Statoil, the revenue accrued to the company, as distinct from the state. The carried interest percentage under the 1969 licences ranged from 5 to 40 per cent.

The licences provided for negotiation of detailed participation agreements. The government had an option to participate, and the right to exercise it became effective on the day on which written notice was served declaring a discovery to be commercial. The government had one year in which to decide. Until the exercise of the option, the government had no right to involve itself in the management, but only to exercise regulatory powers and was entitled to receive information from the company. Once it had exercised its option the government became actively involved in the management together with the other participants. The government also incurred the liability to pay its participants' share of exploration expenses, usually financed out of the government's share of production. In the case of development and production expenses, the government was expected to make its proportionate contribution, though there are instances where the government required the foreign company to meet the government's share of development expenses, to be reimbursed over time.

The policy review of 1971, in Norway following upon the substantial discovery of the Ekofisk field, led to a reinforcement of the policy of participation, and to the creation of a national company, Statoil, on the grounds that effective state participation could only be achieved through such an instrumentality. The 1972 decree expressly provides that the state may, as a condition of granting a production licence, require state participation. The extent of participation was to be negotiated in each case. It is noteworthy that the Ekofisk licence had no participation provision. Following the policy review, a number of amendments were made to the statute governing petroleum operations. The 1972 amendments replaced a flat rate 10 per cent royalty under the 1965 law by a sliding-scale royalty, varying from 8 per cent on fields with production of less than 40,000 barrels a day to 16 per cent on productions of more than 350,000 barrels a day. Area rentals were substantially increased and payment of a production bonus was required. Area-reduction terms which, under the 1965 decree, required 25 per cent of the total area to be surrendered at the end of 3 years and another 25 per cent at the end of nine years, were now altered so as to require relinquishment of 50 per cent at the end of 6 years.

With increasing bargaining power, the terms have been further hardened by Norway. In the most recent licences, a new participation formula has been devised, so that Statoil, the national company, is assured of a 50 per cent interest from the outset. Statoil retains an option to acquire a further percentage, to be exercised after a production profile has been prepared after a commercial discovery, so that it is entitled to acquire (on a sliding scale) up to 75 per cent interest.[72] Also under this agreement, Statoil is not required to reimburse its foreign partners for the costs of exploration. The carried interest formula has thus worked well in Norway − and has been steadily strengthened.[73]

Evaluation of state participation in joint ventures

Governments have favoured state participation in petroleum development ventures as a means for securing the following objectives: (a) increased government take; (b) control over operations; (c) acquisition of technical and managerial expertise; (d) direct access to external markets.

(a) *Increased government take.* Participation means that, in addition to royalty, income tax and any other payments which are to be made by the company, the government also becomes entitled to a share in the net profits. Thus, one economist has characterised participation as "an ingenious way of further increasing the tax per barrel without touching either posted prices or nominal tax rates".[73] This should not, however, lead to the view that this is the only, or indeed, even the dominant objective which a government seeks to secure through participation. For to do so leads to the position urged by some analysts that "government take" can be increased without resort to participation. The arguments advanced by them are that companies prefer to yield a larger government take through increased taxes, rather than through the mechanism of participation and that participation involves the making of investments by government.[74] The fact, however, is that governments attach importance to the other objectives referred to above, and, therefore, participation should not be viewed, or its efficiency assessed, only as a mechanism for increasing government take.

(b) *Control over operations.* Governments accord high priority to control over operations. In response to a questionnaire circulated to some selected governments, a majority emphasised control over operations as the most important consideration which weighed with them in devising their petroleum development arrangements.[75] Some have argued that adequate control is possible through exercise of regulatory powers.[76] The view has also been expressed that[77]

> equity sharing or "participation", may or may not bring the government an effective voice in management decisions within the operating company and may or may not mean that the government plays an active role in other activities leading to the ultimate development of the resource.

Thus, on one view "participation" was a "great non-event" and adverting to the question: "just what does participation give a developing country?" it was observed that:[78]

> It might not give actual participation in transportation; it might not give participation in marketing. It might simply give a host country

a share of the income which the country might have had anyway under the profit-sharing arrangements . . . In particular, it is important to observe that, even if the primary goal of participation is host government representation on the board of directors, in many instances that representation may be virtually meaningless. As outside directors, government representatives are unable to ask the right sort of questions within the board to put forward a rational government policy.

These criticisms are premised on the lack of expertise on the part of governments, and the point is made that government representatives, who have not the requisite competence or experience, are unable effectively to exercise powers of management *jointly* with company representatives, so that they would either be obstructive and, therefore, hamper efficient operations, or would tend simply to endorse the views and decisions of the company representatives, and thus be ineffectual. While no doubt this point has substance, it cannot make out a case for control solely through exercise of regulatory powers. For it is recognised that[79]

> no matter how stringently or extensively applied the control (regulatory) from *without* is it is not enough. Control from *within* is . . . essential. State participation can provide that control in a spontaneous and uninterrupted manner. It entitles the state to have greater representation on the board of management of the operating company, enabling it to play a more effective role in directing the general course of the enterprise and in shaping its policy decisions in both the short and the long-term. Such control would have far reaching influence on such vital issues as the volume of production, pricing policies, the role of investment for further exploration and development, the role the oil industry should play in the national economy, conservation policies and many other related matters . . .

If lack of expertise impairs effective control under state participation, it does so equally in the case of control through regulation. Indeed, the need for expertise in the latter case is the greater, since the regulatory agency is not as closely involved with operations nor does it have the same access to technical and financial information as government representatives involved in a venture in which the government is a participant. Thus, it has been observed that:[80]

> Where the interests of oil companies and governments are parallel, the most extensive regulation of all aspects of operations (including the rate of production), is possible . . . where their interests are adverse, expertise is useful to the government.

The need to gain expertise is, in fact, an argument for participation.

(c) *Acquisition of technical and managerial skills.* It is through direct involvement in each phase, and each level of operations, that expertise is acquired. There is no better method for gaining expertise than on the job, through "learning by doing". State participation provides a basis for involving persons right from the top levels of management to the operational field level. The need to develop a national cadre at each level is essential not only for effective supervision and protection of the state's interest in on-going operations but is an important strategic objective both in terms of progressively reducing dependence on the companies and later when the period covered by the agreement with the company is due to expire, to "take-over" the operations from the company. The experience of Venezuela is instructive with regard to a situation, where the end of the term of the agreement is imminent. The point is emphatically made that in this context companies understandably were unwilling to make needed investments and were inclined to re-deploy technical personnel to other areas.[81] In this situation, the only way in which the state can protect its interest is progressively to assume greater responsibility, which it can only do if it has in the meantime built up its own technical and managerial capabilities.

(d) *Direct Access to External Markets.* Governments continue to have a high degree of dependence on the companies for the marketing of oil. In participation arrangements, under which governments have the right to dispose of their share of the oil, they are provided with an opportunity to enter the world market and explore new outlets. This not only gives them a greater understanding of the market, which is essential if they are to maintain the marketing operations conducted by the companies, but also opens up the possibility of direct sales to certain consumers, without the intermediation of the companies. Thus, Iran's agreement for sale of crude to India, and by Venezuela to different state companies in Latin America, provide examples of a development which marks the beginning of a process of reducing dependence on the companies for marketing.

PRODUCTION SHARING CONTRACTS

The standard Indonesian production-sharing contract

The production-sharing contract ("PSC") was a new type of petroleum development agreement which was pioneered in Indonesia in 1967, and later adopted by a number of countries in South and South-East Asia and beyond.[82]

The form this agreement took can be explained as a response to the requirement of a law promulgated in Indonesia in 1960, Law No. 44, Article 3, of which provided that: "the mining of oil and gas should only be undertaken by the state and that mining undertakings of mineral oil and gas oil are [to be] exclusively carried out by state enterprises". Article 6 of the law, however, authorised the Minister of Mines to appoint foreign companies "as contractors for the state enterprises" if they were required to carry out operations which "cannot or cannot yet be executed by the state enterprise involved".

Under a PSC the contractor bore the risk of exploration, so that if there is no commercial discovery, the loss is borne by the contractor. In the event of a commercial discovery, the contractor is entitled to be reimbursed out of a percentage of the oil produced (referred to as "cost oil") and further, by way of compensation for the work done by it, the contractor is entitled to share in the remainder of the oil (referred to in this context as "profit oil"). The percentage reserved for reimbursement of costs was (until the changes effected post-1973) 40 per cent in each year till all costs were reimbursed and the remainder was split in the ratio 65:35 (65 to the state or the national company, 35 to the foreign contractor); this ratio has been altered post-1973 so that the recent agreements provide for a higher proportion in favour of the national company.

Ownership of the petroleum discovered remained vested in the state (or the national company)[83] and the foreign company, as "the contractor", did not acquire title to its share of the crude oil until it reached the point of export. It has been said that "by merely postponing the title transfer, however, PSCs create an artificial distinction which exalts form over substance" and that "perhaps the only significance of postponing the transfer of title is that it might inhibit legal actions brought by contractors against purchasers of nationalized oil".[84] Attention has been drawn to certain anomalies such as the clause in the contract which requires the contractors to supply the domestic market with crude oil for which it is paid a price (albeit substantially lower than the export price). The question is raised: if the title to the oil does not pass to the contractor till the point of export, then how could it be entitled to receive a price for the oil supplied by it to the domestic market?[85] A question also arose as to whether the contractors, which were US oil companies, could claim the federal depletion allowance in a case where they did not own the petroleum, or have an equity interest in the petroleum operations. This issue was, however, resolved in favour of the companies by a ruling from the Inland Revenue Service, which is said to have relied on a clause in the contract which stated: "The Contractor shall carry the risk of operating

139

costs required in carrying out operations *and should therefore have an economic interest in the development of petroleum operations in the Contract area.*"[86]

The management provisions in the PSC were regarded in the late sixties, when it was introduced, as "a radical departure from conventional arrangements existing in other parts of the world", for it asserted that "Pertamina (the national company) shall have and be responsible for the management of the operations contemplated", while the contractor would be responsible to Pertamina "for the execution of such operations". It was further provided that "Pertamina shall assist and consult with the Contractor with a view to the fact that he (the Contractor) is responsible for the work programme". Formally the national company was vested with plenary powers of management. A closer examination of the structure of the provisions and their actual workings, however, reveals that "the increased *de jure* authority has not necessarily affected the extent of *de facto* control over companies".[87]

The management framework under a PSC is so designed that work programmes in each phase of operations are drawn up by the contractor, and submitted for *approval* by the national company. There is also provision for periodic reporting and submission of information. This type of framework is to be contrasted to one where the national company is *jointly* managing operations with the company, through the instrumentality of a joint management committee; while in the PSC framework, the national company is called upon to approve programmes prepared by the contractors.

An examination of the provisions relating to the exploration phase suggests that in practice they are weaker than they might appear to be. A PSC left it to the company, that is the contractor, to prepare a work programme for approval by Pertamina each year. These programmes were to be submitted *after* the signature of the contract. The only requirements were that exploration work should start within 6 months and that minimum levels of expenditure to be incurred for exploration work each year should be stipulated. It was further provided that if Pertamina proposed a revision to certain specific features of the work programme, it must do so within 30 days after receipt thereof, failing which the work programme would be deemed to have been approved. The weakness of such a provision is manifest. First, under this arrangement, exclusive exploration (and exploitation) rights were granted to the company, before it had committed itself to a work programme; its only commitment being to incur certain minimum expenditures. The expenditure levels (being presented as minimum levels) could prove to be inadequate. Further, the

140

amount of work that a level of expenditure appeared to assure could be further reduced as a result of inflation and/or escalation in the costs of exploration, or if expenditures were incurred which were not strictly necessary. It is usually difficult for a government of a national oil company to monitor the costs being incurred. The provision regarding submission of work programmes for approval which allowed only 30 days to the government or the national company to propose amendments in effect left little scope for the government or national company to propose revisions, since, given the paucity of expertise available to it, it could hardly be expected within such a short time to propose effective amendments. Nor was it possible within the limited time available to seek the assistance of independent consultants.

Under the standard PSCs, companies only submitted the rawest form of data. They declined to submit evaluation reports on the grounds that they were obliged to submit copies only of *original* geological, geophysical and other data and that the evaluation reports did not constitute original data but were "derivative" in character, and further that the cost involved in preparing their reports was not included in the "operating costs". This manifestly weak provision not only denied the national company evaluation reports but further, by not stipulating the types of data or information which the company should be required to collect and submit, left the obligation undefined. This may be contrasted to far more detailed provisions, regarding the data and information which are to be furnished, made in such agreements as, for example, the agreement between Angola and the Sun Oil Group of 1974 (in the concession form), which enumerates the different types of data which are to be collected and submitted to the national company, or in the regulations framed under general legislation regimes such as that of New Zealand.[88] Later PSCs, and in particular those adopted in other countries, have made express provision for submission of evaluation reports, and some have spelt out specifically the important types of data which the company should be required to submit to the government or the national company.

The area reduction/relinquishment provisions varied from contract to contract within a general format, which required a percentage, ranging from 15 to 25 per cent to be relinquished at the end of the third year, another 25 per cent by the end of the sixth year, and in the event of no commercial discovery being made by the tenth year, the entire area was to be surrendered. If, however, there was a commercial discovery, the contractor was entitled to retain not only the area where discoveries had been made, but a substantial area in addition. Thus, the surrender provisions are so formulated as to require surrender of "a portion in excess of . . .", the

141

numbers varying from 5000 to 15,000 square kilometres, or a stipulated percentage of the original contract area.

The provision with regard to whether a discovery could be regarded as commercially significant so as to oblige the contractor to undertake development in the standard PSC was formulated in the following terms:[89]

> If Petroleum is discovered in any portion of the Contract Area within ten (10) years as from the Effective Date, which in the judgement of the Contractor after consultation with Pertamina can be produced commercially, based on consideration of all pertinent operating and financial data such as the size and location of the reserves, the depths and number of wells required to be drilled and transport and terminal facilities needed to exploit the reserves, then as to the particular portion of the Contract Area Contractor will commence development. On other portions of the Contract Area exploration may continue concurrently without prejudice to the provisions of Section 3 regarding the exclusion of areas.

Under this provision, it was left to the judgment of the contractor to determine whether a discovery was commercially significant or not, though in forming this judgment the contractor was required to consult Pertamina and to form its judgment on the basis of operating and financial data. An effective sanction was available in a situation where the issue arose at a point of time when only a single discovery had been made and the time allowed for exploration was due to expire. In such a situation the contractor would be obliged to surrender the entire area unless the discovery was characterised as commercially significant. In situations, however, where a contractor had already made a commercially significant discovery, and the question of commercial significance had to be decided in respect of a subsequent discovery, this sanction would hardly operate, since having regard to the character of the surrender provisions (which allowed retention of a substantial area in addition to the area where a commercial discovery had been made), the contractor could retain the area where the subsequent discovery had been made without assuming the obligation to develop. In this latter type of situation, the national company could find itself in a situation where it could neither require a contractor to develop a particular discovery, nor to relinquish the area in which the discovery had been made. That this is an area of potential conflict between the national company and the contractor is evident from a reported case where Pertamina required a contractor to commence production, despite the contractor's view that on the available information the discovery was not commercially significant. It is reported that the company started development work and incurred certain expenditures, but eventually persuaded Pertamina that it was not feasible to proceed to

142

production.[90]

In the production phase, the contractor is required to obtain approval of its budgets and work programmes. A study of the actual working of PSCs, however, records the finding that[91]

> Pertamina seems fully cognizant of its lack of technological and managerial skills, and aware that it is not in a position to substitute intelligently its own views on petroleum exploitation for those of the admittedly more technically competent contractors. Its authority to give or withold approval is therefore, virtually useless and potentially self destructive.

It is observed that "in the short run, Pertamina has adopted the more sensible goal of educating itself and has not sought to exercise powers which it realises to be beyond its present capabilities . . ."[92] In the judgment of oil and banking executives with experience of Middle Eastern operations the Indonesians, under PSCs "exercise less real control over petroleum operations than do Middle Eastern or North African nationals".[93] The present value of the management clauses is thus seen as "the accretion of technological and mangerial skills gained through constant exposure to the inner workings of the companies".[94] It is questionable, however, whether nationals operating under the management framework under PSCs, which is built around "approvals" being given to programmes prepared by the contractor, have greater opportunities to learn the inner workings of the companies than under joint management frameworks, where the nationals are involved in the preparation of the programmes. It can be argued that there are greater opportunities to learn in the latter case, where the nationals are actually involved at every level of operations.

Under most PSCs, the contractor undertakes to market all the crude oil produced in the contract area, though Pertamina retains the right, by giving prior notice to the contractor to take its share of the crude oil in kind. In a number of the contracts, the marketing obligation of the contractor is reduced and restricted to a quantity to be determined by a formula, which defines the quantity to be marketed by reference to the quantity which the contractor is required to supply to the domestic market. The earliest PSCs did not contain an obligation to supply the domestic market, which was introduced later, and required the contractor, in cases where commercial production had been commenced, to supply a portion of the needs of the Indonesian domestic market, at a nominal price of cost plus 20 cents per barrel.

The problem of pricing arises in a somewhat different form under PSCs than it does under traditional concessions, where the taxes and other payments to be made to the government could be reduced by sales to affiliates at deflated prices. While this type of pricing problem is avoided under a PSC, where "the government take" consists of a share in the production,[95]

> . . . prices nevertheless remain relevant under these contracts. Low prices increase the quantity of oil necessary for the contractor to recover operating costs; thus the total amount of oil available to Pertamina is consequently diminished. Low prices also lengthen the period of time necessary for the recovery of costs, thereby postponing Pertamina's contractual right to larger portions of production.

Further, the contractor can influence the market by the prices at which he sells the cost oil and his share of profit oil, for if he sells them cheaply, this would tend to push down the price, at which Pertamina might be able to sell its share.

The PSC embodies three mechanisms to protect the state's (or the national company's) interests, in the matter of pricing. First, there is a clause which provides that if Pertamina is able to secure a higher price than the contractor for the "cost" oil, the contractor must either match the price obtainable by Pertamina, or permit Pertamina to sell the oil on the contractor's behalf. The proximity to the Janapese market and the marketing apparatus which Pertamina has been able to develop makes this an effective mechanism for ensuring that cost oil is not sold at artificially deflated prices by the contractor. The effectiveness of this mechanism would, however, be substantially diminished where a national company operating under a PSC did not have its marketing apparatus or access to marketing outlets, as would be true of many of the national companies in developing countries.

Secondly, valuation of cost oil sold to affiliates was to be done on the basis of a weighted average per barrel net realised price obtained in arm's length sales to independent third parties, or in sales effected by Pertamina in exercise of its right to sell where it could obtain a price higher than the contractor. In case there were no such sales on the basis of which such a weighted average could be computed, it was provided that[96]

> the value of such sales to Associated Companies [affiliates] should be determined in a commercial manner, taking into account prices at which comparable types and quantities have been sold in competing export markets, bearing in mind in that connection possible differences in quality and in transport costs.

Thirdly, there was an express provision prohibiting the contractor from giving any discount, commission or brokerage to an affiliate, while requiring that any commission or brokerage paid in connection with

144

sales to third parties "should not exceed the customary and prevailing rate". The efficacy of this mechanism "depends largely on Pertamina's ability to identify affiliate transactions".[97]

A further safeguard with regard to financial matters is provided by the power vested in Pertamina to approve the budgets and to undertake auditing and accounting of the operations. The efficacy of these mechanisms depends largely on the capabilities of the agencies which are entrusted with the exercise of these powers. It has been noted that[98]

> Slippage in the amount of income accruing to the government could occur in the calculation of these "operating costs" incurred by the company under post-1965 agreements. Such deductions must be given the quality of scrutiny that would be given by a government tax office to deductions from gross income in a traditional concession agreement.

Such scrutiny requires a high level of sophistication and expertise, which is rarely available to governments, in particular those of developing countries. Even in Norway, the view was expressed that the governmental machinery was handicapped in monitoring the accounts, and in particular, the operating expenses of the companies.[99] Some recent agreements, which may be regarded as advanced versions of a PSC, have incorporated new mechanisms, which will be considered below, for exercising stricter control over "operating expenses".

A more fundamental weakness in the standard PSC's financial provisions is that under them the contractor can earn substantial windfalls in the event of a rise in oil prices; thus, it is observed:[100]

> Although price setting disputes between the parties seem to be successfully avoided by the production sharing formula, the astronomical rise in the price of crude oil precipitated by the Yom Kippur War reveals an inherent weakness of PSC's. As prices outstrip production sharing schemes received a windfall in unexpectedly high profits. When the per barrel price of Indonesian oil was $5.00, production sharing contractors retained $1.05 in "profit". As the price per barrel escalated to $13.00, the Contractor's "profit" increased to $2.75 per barrel.

In the post-1973 situation, the absence of any mechanism in the contract to deal with such a windfall led to the adoption of certain unilateral measures, some of which have been referred to earlier, which included: alteration of the split of 65:35, extending the period of cost recovery, and the application of a "two tier system" for division of oil revenues, whereby the 65:35 split in favour of Pertamina applied only to the initial $5.00 received per barrel, while all monies realised above the

$5.00 ceiling was split 90:10 in Pertamina's favour.[101]

Thus, in 1974, the PSC concluded by Pertamina with Phillips-Tenneco provided that cost recovery would be allowed from only 35 per cent of the production, and the remaining output would be split 72.5:27.5 in favour of Pertamina up to 50,000 barrels per day, 77.5:22.5 per cent when the output was between 50,000 and 150,000 barrels per day, and 80:20 per cent when it exceeded over 150,000 barrels per day. A mechanism to deal with "excess profits" was also incorporated, whereby Pertamina would receive a percentage of the revenues earned from the sale of the profit oil when the sale price exceeded $5.83 per barrel, so that Pertamina would get 85 per cent of the excess revenue, in respect of the first 150,000 barrels per day of profit oil produced, 90 per cent in respect of the next 100,000 barrels per day, and 95 per cent in respect of barrels produced in excess of 250,000 barrels per day.

A further problem adversely affecting US companies which has arisen in relation to the financial package contained in a PSC is the result of a recent ruling given by the Internal Revenue Service of the United States to the effect that a US company is not entitled to a foreign tax credit in respect of certain payments made by Pertamina (partly in satisfaction of taxes payable by the company) to the Indonesian treasury.[102] The ruling was given in the context of a production-sharing contract, proposed to be entered into by Mobil with Pertamina, under which Mobil would recover its operating costs in barrels of oil, but not to exceed an amount equal to 40 per cent of the value of the total number of barrels produced from the contract area during the year. The number of barrels to which Mobil would be entitled for recovery of operating costs was to be determined by dividing Mobil's total annual operating costs by the "weighted average price" of all crude oil produced from the contract area during the year. Except for interest on money borrowed for petroleum operations and the signature production bonuses, paid by Mobil, all expenses incurred by Mobil in petroleum operations were included in the term, "operating costs". The remaining 60 per cent of the oil was to be divided between Pertamina and Mobil: if production was less than 200,000 barrels daily, Mobil would be "entitled to take and receive" 30 per cent and Pertamina 70 per cent; if production were to exceed 300,000 barrels, Pertamina would be entitled to take 72.5 per cent of the barrels in excess of 300,000. Under the contract Pertamina was obliged to discharge the following Indonesian taxes payable by Mobil: transfer tax, certain import and export duties and "exactions in respect of property, capital, net wealth, operations or transactions including any tax or levy on or in connection with operations performed by the contractor", while it was specifically provided that Mobil would remain subject to Indonesian income tax; Pertamina, however, was to pay Mobil's income tax out of Pertamina's share of production. Under the Indonesian law governing Pertamina, it was required to pay 60 per cent of net operating income to the Indonesian treasury, upon payment of which both Pertamina and Mobil would be relieved of any further payment by way of corporate income tax. Mobil had contended that of the net operating income, 31.82 per cent represented royalty payable by it (10 per cent to Pertamina, 21.82 to the government), and 38.18 per cent represented Indonesian income tax and witholding tax, while the remaining 30 per cent represented Mobil's after-tax income. According to Mobil, it should have been entitled to a tax credit in respect of the amount paid to the treasury by Pertamina, which could be said to represent Indonesian income tax. The Internal Revenue Service, however, held that Pertamina's share of production was in substance, if not in form, a royalty and that no part of such royalty was identifiable as an income tax or tax in lieu of income tax and, therefore, Mobil could not be entitled to a foreign tax credit (under Sections 901 and 903 of the Internal Revenue Code of 1943) nor to a deduction (under Section 164), in respect of any part of Pertamina's share of production (applied towards payment of taxes to the Indonesian treasury).

The IRS ruling was followed by a general decision of the IRS, which defined its position with regard to tax deductions in production-sharing ventures, in the following terms: [103]

If a foreign government owns mineral resources and the taxpayer has an interest in such minerals in place, a foreign tax will not be recognised as a tax for U.S. Federal income tax purposes, unless that government also requires payment of an appropriate royalty or other consideration for the payment that is commensurate with the value of the concession. Such royalty or other consideration must be calculated separately and independently of the foreign tax. Satisfaction of such royalty by the U.S. tax-payer must be independent of any foreign tax liability . . .

In order for foreign tax to be credited under Section 901, it must qualify in substance and form as a U.S. tax under U.S. concepts. Generally, in the absence of other factors which have contrary implications, payments to a foreign government owning the minerals in place extracted by the U.S. tax-payer will be treated as a creditable income tax if all of the following characteristics are present:

(1) The amount of income tax is calculated separately and inde-pendently of the amount of the royalty and of any other tax or charge imposed by the foreign government.

(2) Under the foreign income tax and in its actual administration the income tax is imposed on the receipt of income by the tax-payer and such income is determined on the basis of arm's length amounts.

Further, these receipts are actually realized in a manner consistent with U.S. income taxation principles.

(3) The tax-payer's income tax liability cannot be discharged from property owned by the foreign government.

(4) The foreign income tax liability, if any, is computed on the basis of the tax-payer's entire extractive operations within the foreign country.

(5) While the foreign tax base need not be identical or nearly identical to the U.S. tax base, the tax-payer, in computing the income subject to the foreign income tax, is allowed to deduct, without limitation, the significant expenses paid or incurred by the tax-payer. Reasonable limitations on the recovery of capital expenditures are acceptable.

The impact of this ruling on US companies operating under a PSC framework was that it would reduce the net returns to them, as the advantage to the company of the clause under which Pertamina would pay the company's income tax out of Pertamina's share of production would be denied to it. Consequently, the initial reaction of the companies has been to attempt a reformulation of the relevant provisions so as to avail of the benefit of the foreign tax credit, even if this meant that the company would have to pay taxes directly to the Indonesian treasury.[104]

A feature of the standard PSC which made it particularly attractive to the companies was the quick pay-out which was possible under it, so that the company could appropriate up to 40 per cent of the oil towards reimbursement of exploration, development and other operating costs incurred by it. This enabled companies to recover the costs incurred by them within 3-5 years.[105] The form of PSC proposed by Pertamina in 1976 contained a new set of provisions relating to cost recovery. While recovery was to be made from a share of the crude oil produced, it was to be spread out over a stipulated number of years and therefore the companies could not secure payment over a shorter period as was possible when they could appropriate in each year up to 40 per cent of the oil produced. Under the 1976 form it was proposed, in respect of costs incurred prior to commercial production, that all non-capital costs would be allowed as recoverable in equal instalments over 10 years (though in the case of reserves with an estimated life of 7 years or less, recovery could be made over 5 years), and all capital costs would be capitalised as of the beginning of the year of commercial production and depreciated over 14 years. Depreciation was to be on a double declining-balance basis with each year's depreciation allowed as recoverable in each respective year and interest at 8 per cent was payable by Pertamina on the scheduled unamortised non-capital costs.[106] The companies reacted adversely to this change and undertook hardly any new exploration, thereby making it clear that without modifications, the new form of PSC would not be acceptable to them.

The new PSC has retained signature and production bonuses, the former payable by the company to Pertamina at the time of signature and the latter when certain levels of production are attained. These bonuses were not to be treated as part of "operating costs" and were, therefore, not recoverable by the company.

A participation provision, which under some of the recent PSCs reserved an option for a fixed percentage (around 10 per cent) of undivided interest in the venture for an Indonesian entity to be nominated by Pertamina, is incorporated in the 1976 form, which provides that the option must be exercised within 3 months of the notification of the first commercial discovery. The Indonesian participant was to reimburse the contractor for a proportion, equivalent to the percentage acquired, of the exploration and development costs borne by the contractor. This could be paid in oil or in kind by applying up to 50 per cent of the participant's share of the crude oil; in the latter event, payment would have to be made of an amount equivalent to 150 per cent of the amount payable by the participant.

PSCs contain specific provisions requiring the contractor to give preference, in the course of operations, to utilisation of goods and services produced in Indonesia or rendered by Indonesian nationals "provided such goods and services are offered at equally advantageous conditions with regard to quality, price, availability at the time and in the quantities required", and further to undertake "the schooling and training of Indonesian personnel for all positions including administrative and executive management positions", the cost of such training to be treated as part of "operating costs".

The 1976 form also introduces a new provision whereby the contractor undertakes to process up to 10 per cent of its share of the oil in Indonesia and if necessary to establish a refinery for that purpose.

Production-sharing contracts with special features

While in Indonesia the standard PSC underwent the modifications described above, PSCs were adopted in a number of other countries, but with the incoporation of certain special features. The aim was an "improved" PSC by the inclusion of certain mechanisms which were not contained in the standard PSC, and by providing for a higher proportion of the production to be appropriated by the government or the national company. Among the improved PSCs which merit attention are — (a) the Trinidad-Tobago Model Production Sharing Contract of 1974; (b) Libya-Mobil Agreement of 1974; (c) The Indian ONGC-Reading & Bates Agreement of 24 May 1974; (d) the Bangladesh Agreements with ARCO, Union Oil of California and four other companies of September 1974; (e) the Malaysian Agreements with Shell of 30 November 1976, and with Exxon of 8 December 1976; (f) the Egyptian EGPC-Esso Agreement of 14 December 1974, and EGPC-Amoco Agreement of 24 February 1976.

(a) *The Trinidad-Tobago Model PSC (1974).* Unlike the standard Indonesian PSC which allowed a period of ten years for exploration, the term for exploration in the Trinidad and Tobago form was 6 years. The relinquishment provision is also more strict in that it required any two of the four "sub-blocks" making up the contract area to be surrendered not later than the end of the third year. In the exploration phase, the contractor not only assumed a minimum expenditure obligation but also undertook specifically defined minimum work obligations which included the obligation to drill a minimum number of wells to a certain depth, within a stipulated period. It was provided that if at the end of the first or the second third-year period, the contractor had not fulfilled the minimum expenditure obligations, "one half of any amount by which the said sum may fall short of the relevant minimum expenditure obligation shall be forfeited to Government".

The production-sharing formula differed markedly from the Indonesian PSC. The entire production was divided into three shares — A, B and C. The contractor was entitled to Share "A" — a stipulated percentage (which varied according to production levels attained), and to Share "B", which was equivalent to the "Petroleum profit tax, Unemployment levy, compensation tax or any other taxes or impositions whatsoever incurred upon income and profits related to petroleum operations, which the Contractor in accordance with law must pay to the Government". It was provided that Share "B" would be delivered by the contractor to the government in satisfaction of the contractor's tax liabilities. The remaining portion of the production, Share "C" would accrue to the government, thus, the contractor did not enjoy the same freedom as he did under the Indian PSC of applying up to 40 per cent of the production towards cost recovery, leaving the remainder to be split with the national company. Here the percentage recovered by the contractor was fixed, while the government was assured both of an amount representing the taxes payable by the contractor, as well as the ultimate residue. In principle, the government could increase its share by increasing the rate of taxes payable by the contractor; but the companies before signing these contracts, it

150

appears, had obtained undertakings from the government that tax rates would remain "frozen", and would not be varied during the duration of the contract.[107]

The training obligations assumed by the contractor included a specific obligation to provide $50,000 per year during the first 6 years, and $250,000 per year during the ensuing 10 years for scholarships and training of nationals in skills relating to the petroleum industry.

(b) *Libya-Mobil Agreement of 1974.* Under this agreement, exploration expenditures would be borne by the contractor, and would not be recoverable even if a commercially significant discovery was made. In case of such a discovery, production would be divided 85 per cent to the government and 15 per cent to Mobil in on-shore areas, and 81 per cent to the government and 19 per cent to Mobil in the off-shore areas. A substantial proportion of the development costs would be advanced by the Libyan Government — 85 per cent in the case of on-shore discoveries, and 50 per cent in the case of off-shore discoveries. The advance was to be paid back to the government in twenty annual instalments, with interest (except that 30 per cent of the advance made for development of any off-shore discoveries would be interest-free).

(c) *The Indian-ONGC-Reading & Bates Agreement of 24 May 1974.*[108] The total duration of the agreement was 24 years, but the exploration phase was limited to 7 years (84 months) so that if no commercial discovery was made within that period, the entire contract area would have to be surrendered. The relinquishment programme was more strict than under the standard PSC. By or before 15 months, the contractor would select an area of 5000 sq. km. Exploration phase I would extend over 36 months, and at the end of that period, the contractor would retain not more than 2500 sq. km., though he could retain up to 5000 sq. km. by paying in advance in each quarter a sum of $50,000. By the end of Exploration phase II, that is, by the end of the fifth year, the contractor could retain only 2500 sq. km. If at the end of phase I or phase II a commercial discovery had been made, the contractor could retain an area up to 1250 sq. km. for up to 4 years if he undertook to execute an agreed exploration work programme.

Further in the exploration phase, in addition to a minimum expenditure obligation, a minimum work programme was to be undertaken. The type of data to be furnished was spelt out in some detail. Upon a discovery being made, the question whether it was commercially significant or not was to be determined by the "mutual judgement of ONGC and the Contractor". Of particular note is the introduction of a "sole risk" clause under which if the contractor did not consider the discovery to be commercial and ONGC held a different view, the contractor was obliged to relinquish the portion of the area required for development by ONGC, with the contractor having no rights in the petroleum produced from the area.

The instrument of a joint committee of six, three representing ONGC and three the contractor, was introduced to exercise continuing supervision over operations, including in particular: (a) review of reports on operations, (b) consideration of annual work programmes and budgets, and any major revisions thereof, (c) consideration, in case of discovery of proposals for appraisal and development of such discovery and (d) generally reviewing and supervising the implementation of the respective obligations by the partners. The committee was expected to meet at least quarterly.

The available data on the financial terms and production shares indicated that cost recovery could be effected from up to 40 per cent of the production, and the remainder would be split 65 for ONGC and 35 per cent for the contractor, though the government's share could increase to 80 per cent at specified levels of production.[109] ONGC would have the first option on all crude oil produced to meet internal requirements, at prices based on Indonesian and Gulf crudes, with the contractor being allowed to export any surplus once production was sufficient to meet internal requirements.

Provision was also made for the government to have an option to acquire 10 per cent equity interest upon a commercial discovery being made. A less-publicised, but stringent, option, is reportedly included in the agreement under which ONGC has the option to acquire the entire interest once the contractor has earned three times the total amount invested by it.[110]

(d) *The Bangladesh PSCs (September 1974).* The duration of the contract was 21 years. The relinquishment provisions were so structured that 25 per cent was to be relinquished by the end of the third year, a further 25 per cent by the end of the sixth year, the total area to be reduced to an area not larger than 1000 square miles by the end of the eighth year (exclusive of the area covered by commercially exploitable fields). At the end of the eighth year all areas were to be surrendered except any area covered by a commercially exploitable field (though there is a proviso, for retention in certain circumstances, of an area not exceeding 1000 square miles up to the end of the twelfth year).

In the exploration phase, in addition to the minimum expenditure obligations, there was an obligation to commence drilling not later than

24 months from the date of the signing of the contract. In addition to submission of annual work programmes, the contractor was required to submit proposals of work to be undertaken in the coming 2 years. The provision requiring submission of information and data to the national company, Petrobangla, expressly required that the contractor should submit along with original data "all analyses and interpretations thereof".

The production-sharing formula was structured in a sliding-scale form based on the levels of production attained. Thus, if the production was up to 50,000 barrels per day, the contractor would be entitled to recover operating costs of an amount equal in value to a maximum of 30 per cent of the total production, if production exceeded 50,000 barrels per day but not 100,000 barrels per day, the percentage was reduced to 25, and if it exceeded 100,000 barrels per day, then the percentage would be 20.

The division of profit oil would also be on a sliding scale basis, thus:

		Petrobangla	Contractor
If production was	100,000 barrels	76%	24%
For that part in excess of	100,000 and not exceeding 150,000	77½%	22½%
For that part in excess of	150,000 and not exceeding 200,000	80%	20%
For that part in excess of	200,000 and not exceeding 500,000	85%	15%
For that part in excess of	500,000 and not exceeding 700,000	87½%	12½%
For that part in excess of	700,000	90%	10%

Provision was made for payment of production bonuses also on a sliding scale related to levels of production attained. In addition, the contractor would pay 3 cents per barrel on the barrels recovered in excess of 100,000 barrels as a contribution towards research related to petroleum or any other activity as may be determined by Petrobangla. Further, in pursuance of its obligations to train nationals, the contractor would contribute towards the cost of training to the extent of $100,000 per year.

(e) *The Malaysian PSCs with Shell (30 November 1976) and Exxon (8 December 1976)*. The Malaysian PSCs followed the enactment of the Petroleum Development Act, 1974, whereby the national oil company, Petronas, was given exclusive rights for exploration and exploitation of petroleum. Petronas was empowered to engage foreign companies as contractors, and existing concessions were required to be converted into PSCs with Petronas. The initial reaction of such existing concessionaires as Exxon and Conoco was to announce a temporary suspension of activities. A provisional production-sharing formula agreed upon in December 1975 gave 7.5 per cent of the excess revenues from crude oil sales to the foreign company.

The PSCs ultimately concluded with Shell (3 November 1976) and Exxon (8 December 1976) embodied a sharing formula significantly weighted in favour of Petronas. The duration of the agreement was 20 years, with a possible extension of 4 years for oil, and 14 years for gas. For cost recovery, the contractor would be permitted to retain up to 20 per cent of the total production in the case of oil, and 25 per cent in the case of gas. A royalty equivalent to 10 per cent of the production would be realised. The balance would be split 70 (for Petronas) and 30 (for the contractor), with the obligation that both Petronas and the contractor would pay income tax separately on their respective shares. The income tax was reduced from 50 to 45 per cent in late 1976 — the 10 per cent royalty, bonus and research contribution (½ per cent the contractor's sale proceeds on both profit oil and cost oil) were deductible from the tax. It is estimated that the effective split under the Shell contract was 83.5 (Petronas) and 16.5 (contractor) and under the Exxon contract was 92.5 (Petronas) 7.5 (contractor).

A mechanism to deal with "excess profits" arising from price escalation is provided, so that when sale prices exceed $12.70 per barrel, the contractor would pay 70 per cent of its sale proceeds over the basic price to Petronas (a "normal" increase in price was allowed of 5 per cent per year on the price fixed as at end 1976).

(f) *The Egyptian EGPC-Esso Agreement (14 December 1974) and EGPC-AMOCO Agreement (24 February 1976).* The EGPC-Esso agreement represents a hybrid form where the joint operating company instrument, characteristic of joint ventures, has been grafted on to a PSC. The non-profit joint company was to come into existence within 30 days after a commercial discovery, to undertake all development and production operations. Prior to the coming into existence of a joint company, an exploration advisory committee of six persons, three to be appointed by EGPC and three by Esso, would be consulted and would advise on exploration operations, and work programmes would have to be approved by it.

In the exploration phase, the contractor assumed a minimum expenditure obligation, with the further obligation that it would start off-shore drilling not later than the end of the thirteenth month from

154

the signing of the contract.

Specific mechanisms were incorporated to monitor costs. The contractor was required within 45 days after the end of every quarter to supply EGPC with a list of costs incurred, and necessary supporting documents were to be kept available for inspection. EGPC could within 3 months from the date of receiving such a list, raise an objection on any of the following grounds: (a) that the record of costs and/or proceeds were not correct; (b) that the cost of goods or services supplied were not in line with the international market prices for goods and services of similar quality supplied on similar terms, prevailing at the time such goods or services were supplied; (c) that the condition of the materials furnished by the contractor did not tally with the estimated prices; or (d) that the costs incurred were not required for operations under the agreement in accordance with practices usual in the international petroleum industry. A further check was provided with regard to costs to be incurred in later phases by the requirement that the approval by the Managing Directors of the joint company would be required for (a) major capital expenditures and (b) major contracts for supplies and service. Any disagreements between EGPC and the contractor were to be resolved through consultation.

The production-sharing formula provided for cost recovery from 25 per cent of the production (in waters up to 200 metres in depth) and up to 40 per cent (where the water depth exceeded 200 metres). A recovery rate was stipulated whereby exploration expenditures would be recovered at the rate of 20 per cent per annum, and development as also producing expenditures, with the exception of operating expenses which would be recovered in the year incurred. Profit oil would be split 80/20 in favour of EGPC (where the water depth did not exceed 200 metres) and 70/30 (in water depths exceeding 200 metres).

The contractor was under an obligation to develop a commercially significant discovery, but where a discovery was not deemed by the contractor to be commercially significant within 4 years from the date of the discovery, EGPC would be entitled to develop it at its sole risk. It was provided that if there was a subsequent commercial discovery in that area, the contractor would have an option to share in the development and production of that discovery, after EGPC had recovered 300 per cent of the costs incurred by it, and Esso had made a payment to EGPC equal to 100 per cent of the costs incurred by EGPC.

The EGPC-AMOCO agreement is similar in structure to the EGPC-Esso agreement in that a non-profit joint company was empowered to undertake all development and production operations, while a joint committee supervised exploration operations. Some of the mechanisms embodied in the later agreement have been further strengthened.

Thus, the mechanisms for monitoring costs on the basis of quarterly reports to which objections could be raised in the manner indicated above in the EGPC-Esso contract, were incorporated with the addition of a clause to the effect that if a mutually satisfactory settlement of any accounting or cost objection between EGPC and AMOCO was not reached within 3 months of the raising of the objection, either party could refer the dispute to a firm of chartered accountants of international repute, agreed upon by them, for determination, and failing agreement, by a firm to be nominated by the President of the Institute of Chartered Accountants in England and Wales.

The definition of a "commercial discovery" was simplified. A commercial well was defined as

> the first well or any geological feature which after testing in accordance with sound and accepted international industry production practices, and verified by EGPC, is found to be capable of producing at the average rate of not less than 1000 barrels of crude oil per day, unless a lesser average of tested crude oil production is justified commercially.

A sole risk clause was included under which if the contractor did not deem a discovery to be commercial, EGPC could decide to develop it at its own risk and cost, in which event the area which covered the discovery would be relinquished by the contractors, who would have no rights in respect of that area.

Cost recovery was to be effected from 20 per cent of the crude oil produced, but recovery was to be spread out over a period of time, and different rates are stipulated in respect of different categories of cost. Thus, while operating expenses are recoverable in the tax year in which they are incurred, exploration expenditures are recoverable at the rate of 25 per cent per annum based on amortisation at that rate starting in the year in which such expenditure was incurred and development expenditures at the rate of 12½ per cent per annum. Different rates are also stipulated for "sunk costs" and "warehouse stock value".

Provision is made for valuation of the cost oil at the "market price" to be determined by agreement and, failing agreement, by arbitration.

Evaluation of production-sharing contracts

The advantages of a production-sharing contract from the point of view of oil companies have been noted, thus:

Production-sharing agreements have been popular with the oil companies. The companies control their own share of the crude oil and, barring an election by the state oil company to take its share in kind, they can control the destination of the state oil company's share. Moreover, no tax is payable by the company. Finally, and most importantly, companies have been able, on their share of the crude oil, to enjoy the whole of the price increases in the world market.[111]

The last of the advantages for the companies mentioned, that is the enjoyment of the whole of any price increase, is seen from the government's point of view as the principal weakness of the PSC. The problem of monitoring costs and exercising effective supervision has also been noted. A study of the working of the PSC in Indonesia, while it records that[112]

On the basis of empirical investigation, it is reasonable to conclude that the operational differences between production-sharing and concession contracts are virtually meaningless,

goes on to enumerate what may be regarded as the strengths of the PSC from the point of view of host governments:

The minimal present value of the contractual provision (relating to powers of management) should not eclipse its important psychologically settling effect on the relationship between Pertamina and the contractor. By creating a mechnism for a continuing dialogue between the parties, it has fostered a substantial mutual understanding and confidence . . . however, the management clause satisfies a principal anti-concessionary objection to the extent it provides the means by which Indonesians may eventually gain control of the petroleum sector . . . In a similar vein, the provision that postpones the transfer of title from the well-head to the point of export obviates the concessionary appearance of foreign ownership of Indonesian oil (yet the operational and legal significance of postponing the title transfer appears to be minimal). Contractual clauses that implement price controls and that call for a division of oil rather than profits are significant improvements over the standard concessionary regime, *as they ensure a role for the state enterprise in marketing and eliminate disputes over prices* . . . the important differences which do exist are more the result of the proficiency of Pertamina than of the production sharing contract. The potential value of the contract lies in the numerous provisions which minimize conflict by maximising Pertamina's control at those points at which the interests of the companies are most likely to diverge from those of Indonesia (or Pertamina).

The "improvements" made in subsequent PSCs in Indonesia, and in the PSCs adopted in other countries, have sought to remove some of the weaknesses of the standard PSC. Thus, stricter relinquishment provisions have been included as in Trinidad and Tobago and India. Minimum drilling obligations have been incorporated in the Trinidad and Tobago form, and in India, Egypt and Bangladesh. The Indian and Bangladeshi forms stipulate specific types of data and material which are to be submitted when a discovery has been made. The Indian and the Egyptian forms contain "sole risk" clauses enabling the government or the national company to develop a discovery which a contractor did not regard as commercially significant. The Egyptian form contains specific mechanisms for monitoring and controlling costs. The new Indonesian contracts, as well as the Egyptian contracts seek to regulate the rate at which costs can be recovered. The Indonesian and the Malaysian have both sought to incorporate mechanisms for dealing with the problem of "windfall profits" arising from substantial price increases, and most of the recent PSCs revise the ratio in which the production is to be shared in favour of the government or the national company. The Egyptian agreements have adopted the instrument of a non-profit joint company to enable the national company to play a more active role in management than under the standard PSC.

The PSC has, thus, proved to be a flexible instrument, in which improvements can, and indeed have been made, through the incorporation of additional mechanisms to deal with situations which were not adequately dealt with in the standard PSC. It is important to emphasise, however, that the above analysis indicates that the standard PSC contains substantial weaknesses, so that parties entering into one need to devise and incorporate specific provisions to safeguard different interests which are inadequately secured in a standard PSC.

SERVICE CONTRACTS

"Service contracts" is a description which embraces a range of different types of petroleum agreements. It is important to distinguish the service contracts pioneered in the sixties in the first wave of efforts to replace traditional concessions from the more advanced form developed in the seventies.

The basic features of a service contract have been described thus:[113]

> The national oil company is by law the sole titular holder of the area under agreement. All petroleum deposits and oil and/or gas produced are the property of the national oil company at the well head. The foreign company, either directly or through a subsidiary, acts as general contractor for the national oil company, and as such carries out, in the name and on behalf of the latter, all operations necessary for the exploration and development of oil deposits. Thus, the

contractor is not a concession holder or partner, but merely a hired agent. The foreign contractor is solely responsible for providing all the necessary funds at his own risk. In practice, this means that unless oil is found in commercial quantities, the foreign contractor will not be reimbursed for the expenses he has incurred in his unsuccessful search for oil. However, if oil is found, the cost incurred up to the time of discovery will be considered as a loan to be debited to the national oil company's account. The loan will be repaid in the form of crude oil, after the start of commercial production . . . Once oil is produced, the contractor undertakes to act as an agent for the national company and markets the oil abroad if the latter so wishes in return for a certain nominal commission. The contractor has the right to retain a certain percentage of the sale proceeds in repayment of the loans that he advanced during the exploration and development stages.

The early service contracts are exemplified by the Venezuelan service contracts formulated in the sixties and the ERAP contracts, with NIOC in Iran (1966) and INOC in Iraq (1968). These forms, which are analysed below, should, however, not be confused with a variety of "contracts of work" which were adopted in different countries in the sixties, such as those concluded between 1958 and 1961 by the Frondizi Government in Argentina and the "contracts of work", concluded in 1963 with the major oil companies operating in Indonesia.

The Argentinian contracts of work ("the Frondizi contracts") became the source of major political controversy for while

> Frondizi and the oil companies regarded the contracts as legal service contracts, not concessions, the Ilia government regarded the exploration and development contracts as really concessions in disguise.[114]

The Ilia Government had therefore proceeded to annul these contracts in 1963, by a decree which embodied the reasons why the "contracts of work" were to be regarded as concessions in disguise, thus:[115]

> The legal fiction of works contracts has been utilized to dissemble concessions; however, it is an axiom of our juridical structure that an institution is qualified juridically not by the name which is applied to it but by its own attributes. (Civil Code, Article 1326);

> The works contract, and institution of civil law, is a contract whereby one party (the agent) assumes the obligation to perform certain material or immaterial results for the other party (the principal) for a price in money and without subordination of the agent to the principal (Civil Code, Article 1493);

> The works contract is essentailly bilateral, onerous, mutual, of successive stages and commutative, which signifies that the parties agree

159

upon a fixed or determinable emolument, subject to immediate appreciation of the commutative equivalence established ab initio;

None of the development contracts respond to the indicated juridical qualifications, because they do not concretely specify the works to be performed, the price to be paid, the obligation to verify and to receive such works or the establishment of delivery locations and dates. A determined area has been assigned to each company, which is then left entirely free to act within that area during the life of the contract, and Yacimientos Petroliferos Fiscales has *no faculty to intervene in the management of the development*: on the contrary, the state enterprise is obligated to refrain from restricting production by the contractors and is committed to accept all petroleum produced by them regularly and uninterruptedly; not only is there no specification of work to be delivered at the expiry of the contracts, but also development rights are conceded for excessively long periods, and the emoluments to be received by the contractors for the petroleum delivered are stipulated in kind or on the basis of a determined cost plus a profit calculated as a percentage of the price of a cubic meter of imported petroleum plus insurances and freight, or on the basis of the price of imported petroleum subject to certain fluctuations of international market; [italic wording added].

The pressure on the majors operating in Indonesia to negotiate the replacement of their existing concessions resulted in the conclusion of contracts of work by Shell, Stanvac and Caltex on 25 September 1963. The basic features of these contracts of work were: [116]

They included an undertaking by the contractor to explore and develop diligently, backed by minimum exploration investment commitments, and mandatory relinquishment of 25% of the area after five years of exploration and another 25% after ten years. A sixty-forty profit split in favour of Indonesia was also provided with both a signature bonus and production bonus of $5 million each for the new areas. The "realized price" concept was retained and there was provision for a value committee in the event of dispute as to price. The foreign contractor was appointed as the exclusive sales agent to market the state enterprise's oil, but the state enterprise reserved the right to elect to take 20% of aggregate production in kind. *Management control was retained by the foreign contractor* . . . The contracts also provided for the sale of all Shell's and Stanvac's refineries and the domestic marketing and distribution assets of all three companies. Refining facilities would be sold over a period to begin in ten years and end in fifteen years . . . The companies also agreed to supply the Indonesian domestic market with crude oil and refined products at cost plus fixed fees. [italic wording added]

Thus, while the oil company under these contracts undertook "to return all of its rights to mine mineral oil and gas in Indonesia", it was appointed as "the sole contractor" for a designated state enterprise "to accept the rights and obligations to conduct mineral oil and gas mining operations . . . to provide all financing . . . and technical skills required for the operation . . . [and would] have effective control and management of the operations . . . and full responsibility therefor and assume all risks thereof".

Service contracts were developed in Venezuela on the basis of Minimum Terms formulated in 1968, while ERAP concluded a service (or agency) contract with NIOC in Iran in 1966, and INOC in Iraq in 1968. These may be considered as the "Early Service Contracts" while those which have been developed, in particular since 1973, may be characterised as "Advanced Service Contracts".

The early service contracts — Venezuela, Iran and Iraq

(a) *Venezuela.* With a policy of "no more concessions" being adopted in 1958, the national oil company, Corporacion Venezolana del Petroleo (CVP) was established in 1960. The role of the foreign companies was to be transformed from that of concessionaries owning the oil to that of contractors rendering services. Thus, a "service contract" was concluded with Mobil in September 1962 for the production of oil. The need, however, was felt to devise a new type of contract under which foreign companies could be engaged on a basis where they would bear a part of the risk for exploration but at the same time allow participation in management to CVP, and lower profits to the companies than under concessions. Bases and norms for "service contracts" presented by the National Council of Energy in October 1966 to the Venezuelan Congress set out the main features of the proposed service contract in the following terms:[117]

> The main objectives of the service contracts and mixed companies suggested by the Council were greater financial and operative participation for CVP than had been obtainable under the old concession system. To meet the problem of risk in exploration and development, the new mixed companies formula provided that CVP's contribution of capital was to be made only after production had begun; until that point had been reached, all costs were to be borne by the private partner in the mixed Company. Similarly, in service contracts for development, all capital was to come from the contracting company until the initial period of risk was over, after which CVP would have the option of contributing capital and participating in profits. Additional government revenues were to come from special payments geared either to the rate of profits per barrel on the oil extracted and sold by service contractors, or to the rate of physical productivity of the area under development. Management participation by CVP was

to be established by CVP's membership on the board of directors of the mixed company and by agreement between the two parties in selecting personnel to fill administrative, management, and operational or technical positions in the mixed company. Decisions of special importance by the board of directors were to be reached unanimously, although the board was to be made up of equal numbers of CVP and private company members.

The amendments made to the Hydrocarbons Law in 1967 authorised the employment of foreign oil companies under "service contracts".[118] While conferring authority on the government and CVP to negotiate individual contracts, the 1967 law laid down certain general requirements, namely, that the terms of each new service contract negotiated must be more favourable to the state than that prevailing under the existing system of concessions. This provision was introduced to reassure Congress, and the reason for it appears in the presentation of the proposals made to Congress by the Director-General of CVP, thus:

> One concession does not necessarily have to resemble another . . . A contract may be even more disadvantageous than a concession . . . it is for that reason that it is assured, that it must be assured, that the service Contracts represent evident improvements with respect to the system of concessions . . .[119]

It was further laid down that the new agreements must specify that all lands, permanent structures and equipment forming part of the project would become the property of the state at the termination of the agreement: this provision anticipated a similar provision which was introduced in the standard Indonesian PSC, and is in contrast to the provision in the Indonesian "contracts of work" of 1963, which expressly provide that the contractors were free to dispose of all assets which were used in connection with their operations.

Following the enactment of the legislation, CVP published in March 1968 a set of Minimum Terms for Service Contracts. It was clearly stipulated that:[120]

> 4. Nature of the Contracts — The contract shall be for direct services, whereby the contractor will undertake to adequately explore at his sole expense the area under contract, and to extract the recoverable petroleum, delivering same to CVP, who in turn shall transfer to the contractor a quantity, which shall be stipulated in the contract, and which shall not exceed ninety per cent (90%) of the production, for its sale in foreign markets.
> 4.1 The contractor shall charge CVP the cost of the total production of the area.
> 4.2 The income of the contractor shall be contributed by the receipts derived from the sale of the petroleum received from CVP *less* the

cost of producing the petroleum and other chargeable expenses.

4.3 On the income thus obtained — net profit — the contractor shall pay to the National Treasury the income tax which may be applicable.

4.4 The contractor's ultimate profit shall be finally determined by applying CVP's additional Sharing Scheme according to the Productivity.

The duration of a contract consisted of a 3-year exploration period followed by a 20-year exploration period. A minimum investment obligation and a work programme, including a minimum exploratory drilling programme, would have to be presented by each applicant, along with his bid. If within a period of 3 years the minimum programme was not complete, the contractor would pay CVP the balance of the investments agreed upon. The obligation to develop a discovery would arise "when estimated recoverable reserves" in the exploration area selected by the contractor are such that they reasonably allow an economic projection to be made of an exploitation programme; the mode of making the projection is spelt out in the following terms:[121]

7.1 To establish this projection, income from sales will be calculated at applicable realization prices. The cost of exploration, development, exploitation, taxes, participations and other expenses that may be applicable will be deducted from the year to year income. To the net profits series thus obtained, the corresponding annual depreciation and amortization shall be added. This cash flow series shall be adjusted from year to year against the investments made to obtain an Adjusted Cash Flow Series to which a reasonable discount rate shall be applied as from the moment that the first income can be calculated for petroleum sales of the area. If the result of the algebraical sum of discounted values were zero or a positive amount, the profitability of the production will be considered as feasible.

The mode for selection of the exploitation area by the contractor allowed "alternate choices" to CVP and the contractor, on the basis of which the contractor could select for exploitation blocks comprising 20 per cent of the original contract area.

A basic condition for exploitation was involvement of CVP in all phases of operations, the requirement being stipulated in the following terms:

9.1 In the formulation, execution and supervision of the plans, programs and budgets required for development, it will be kept in mind that one of the fundamental objectives of service contracts is to attain in all phases an ample operating participation of the Nation through CVP, which must be assured through the establishment of Committees composed of representatives of CVP and the contractor.

The contractor undertook to market the quantities of petroleum to which it would be entitled under the contract, at prices which would be fixed through the joint participation of CVP and the contractor, who would follow "commercial norms" for this purpose.

The risk of no discovery was to be borne by the contractor. In the event of a discovery, if commercial production was established, the contractor would receive up to 90 per cent of the petroleum from the sale of which he could recover his investment and earn profits, on which he would be liable to pay income tax.

CVP would have the option to acquire equity participation once commercial production had been established on terms to be agreed.

The congressional approval accorded to the terms of the contract further provided for additional payments to CVP to be based on the contractor's net profits.

The Venezuelan service contract was adopted in 1971 to fill "a void in the petroleum policy that had prevailed throughout the preceding twelve years".[122] The level of exploratory activity by the existing companies was substantially reduced and CVP itself had not been in a position to undertake extensive exploration. The "service contract" form of 1971 was, therefore, formulated to attract oil companies into this field. The rapidly changing environment and in particular the post-1973 developments, accentuated the wariness of the companies with regard to these contracts, so that in all only five such contracts were signed in 1971, three by Occidental, one by Shell and one by Mobil.

A reason given for the poor response on the part of the companies was that:

> Companies had politicized the service contracts to the point where many of the companies were rightly skeptical of the state's long run ability to abide by the new system. Once the issue of "disguised concessions" appeared in public discussions, it did not take a sophisticated forecaster to conclude that the whole arrangement was insecure.[123]

The 1971 service contracts were, thus, seen on all sides as a "temporary and transitory device".[124] Advanced service contracts are at present being formulated on the basis of which the services of oil companies are proposed to be procured for exploration by the new national entity — Petroleos de Venezuela (Petroven).

(b) *Iran and Iraq — ERAP Contracts.* The French state-owned oil company, ERAP, pioneered a type of "service contract", also described as an "agency contract", in Iran in 1966, and in Iraq in 1968.

ERAP was designated "general contractor" under a contract for "technical, financial and commercial services" to be rendered by ERAP and/or

164

its fully owned affiliates and to be reimbursed through the guaranteed sales at an agreed price of 30 per cent of the quantities (of oil) discovered and produced and not set aside as national reserves. The contractor would undertake exploration in the contract area, and in the event of a commercial discovery, would undertake development and production. The funds necessary for these operations would be provided by the contractor. The funds advanced for exploration operations would not be recoverable, if no discovery was made; in the event of a commercial discovery, they would be recoverable as an interest-free loan. Funds advanced for appraisal and development would be recoverable as interest-bearing loans. The contractor would undertake to assist with the marketing of the oil. Such assistance would take the form either of brokerage activities performed by the contractor or purchase by it of minimum quantities of crude oil annually within limits stipulated in the agreement.

It was clearly stipulated that the national company would be the sole owner at the well-head of all petroleum produced. It would also own any land and fixed assets purchased for the purposes of petroleum operations under the contract, as well as movable assets used "in a permanent manner" in those operations, if their total cost had been so charged to the operations as to be recoverable by the contractor.

The duration of the contract consisted of an exploration period of 6 years, followed by an exploitation period of 20 years. The exploration period would cover three phases, the first of 3 years, the second of 2 years, and the third of 1 year. During the exploration period, the contractor was authorised "to manage and conduct exploration and appraisal operations". The contractor was to establish annual work programmes and budgets after "consultation" with the national company. The contractor was subject to a minimum exploration obligation in each phase, and was required to commence drilling of a well within 9 months of the signing of the contract. In case the minimum expenditure obligation was not fulfilled, the unspent balance was to be paid by the contractor to the national company. An area reduction provision required that at the end of the first phase, the contract area would be reduced by 50 per cent, at the end of the second phase the area would be further reduced so that the area retained would be not larger than 25 per cent of the original contract area, and unless there was a commercial discovery, the total area would have to be surrendered at the end of the third phase.

The contractor was obliged within 3 months of the completion of a discovery well to prepare an appraisal programme and budget, after consultation with the national company. A discovery oil well was defined as:

a well of which a production capability of not less than 2000 bbl/d of crude oil for horizons not exceeding 2,500 metres in depth or 3000 bbl/d for deeper horizons [could] be established when tested according to sound oil field practices and under the drive of natural reservoir energies.

At the conclusion of the appraisal programme, the national company could require the contractor to drill further appraisal wells. If the contractor agreed to the proposal, funds would be advanced for this purpose in the usual manner provided under the contract. If the contractor did not agree with the proposal, it would finance the appraisal wells only on terms to be specially negotiated and agreed.

In the exploitation (i.e. production and marketing) period, in the initial phase, management would be conducted by the contractor under the supervision of a joint operating committee, composed of two representatives of the national company, and two of the contractor. This committee would decide the following questions: (a) programmes and budgets (annual as well as quinquennial); (b) operating expenses and personnel; (c) fixation of rates of production; (d) generally, any question likely to lead to significant variations of the costs and/or rates of production; (e) determination of the price of crude oil (including posted price).

The contract further provided that management would be assumed by the national company from a "take over date", which would "occur after five years from the date of commercial production, provided that the last instalment of the development loan had been fully repaid". It was agreed, however, that after the national company had taken over the management, any decision that provoked a major variation of the costs and/or the volume of production would require the contractor's agreement; the contractor would not "unnecessarily" withhold its agreement, and such agreement would not be necessary where it appeared "from the common study of the decision envisaged that it was not likely to provoke a major variation of costs and/or value of production".

The fee, or the remuneration, for the services rendered by the contractor took the form of a guaranteed right to purchase up to 30 per cent of the crude oil deriving from each of the exploitation areas, at a certain formula price which, in effect, allowed a substantial discount to the contractor.

The funds advanced by the contractor for exploration would only be recoverable if a commercial discovery had been made, and the method of repayment of funds advanced for exploration and for appraisal, development and exploitation are separately stipulated in the contract. Exploration funds would be recoverable from the end of the first year of commercial production; the amount recoverable would be the greater of: either one-fifteenth of the total exploration expenditure or the amount

166

arrived at by multiplying 10 cents by the total number of barrels produced. Development loans were to be recovered over a period of 5 years in ten semestrial instalments. It was provided, however, that repayment on all types of loans and interest may be limited should the national company so wish to the "net annual cash flow prior to repayment of loans" for that year, so that the national company would never have to rely upon monies not derived from the cash flow of the operations under the contract to effect repayment of the loans.

The contractor also assumed the obligation to extend assistance to sell certain quantities of crude oil. The contractor could either sell to third parties, itself acting as a broker and receiving a commission per barrel ranging from 0.5 cents to 1.5 cents per barrel, or by purchasing the oil itself at the international market price, less a discount ranging from 0.5 to 1.5 cents per barrel.

In comparing the ERAP contracts (described as "agency-type contracts") to a concession, the following characteristic features have been noted:[125]

> Comparisons can be made on the economic and legal levels. In the legal field the difference can be summed up in two particular points: the ownership of the oil and installations which in concession-type contracts belong to the company, has been recovered from the operator in the agency-type contract . . . From the economic point of view the differences are more notable because each system starts from a completely different conception. In the concession-type contract the company is a contractor in the broad sense of the word, which finances, produces and sells for its own account and only owes to the state the taxes applicable to its activities. By contrast, in the agency-type contracts, the general contractor is separately a financier who loans capital, a broker who sells a part of the production at the market price, and an operator who is paid in part at cost price for his services and in part by means of a right to buy a proportion of the oil produced at an agreed price.

Advanced service contracts

In the post-1973 environment, certain governments, in a strong bargaining position, have developed advanced forms of service contracts, with terms even more advantageous to them than were embodied in the early service contracts. Noteworthy among these are: (a) Burma — Service contracts proposed by Myanma Oil Corporation (1973); (b) Iran — NIOC Service Contracts with Deminex, CFP, Ultramar and others (1974); (c) Brazil — the Model Service Contract proposed by Petrobras (1976).

(a) *Burma — service contracts proposed by Myanma Oil Corporation (1973).* In 1973, the Burmese national oil company notified the "general

167

terms and conditions" of service contracts on the basis of which oil companies were invited to apply for petroleum exploration in off-shore areas. Applicants were required to submit annual exploration programmes together with financial commitments, the latter to be supported by a bank guarantee. The contractor was to bear all exploration risks and to finance subsequent development work. If oil or gas was discovered in the contracted area and if the foreign company did not consider the discovery to be of commercial quantities, then the contract area was to be surrendered to the national company. In the event of a commercial discovery being made, reimbursement of the expenditure incurred by the oil companies for exploration and development would be made out of the sale proceeds of any oil or gas discovered. The distinctive feature of the Burmese service contract, as proposed, was that the remuneration or fee for the services rendered by the contractor would be in the form of a share of *profits* realised from the sale of any oil or gas that was discovered. It was expressly stipulated that the national company would have "management control" in "planning, work programmes, budget, and procurement" and further that all operations would be supervised *jointly*.

(b) *Iran – NIOC service contracts with Deminex, CFP, Ultramar and others (1974).* The six new service contracts entered into in 1974 by NIOC in Iran followed amendments to the petroleum law of 1957 and introduced a new framework for relations with oil companies. Under these contracts, the contractor undertook exploration at its own risk and cost. The exploration period was 5 years. If no discovery was made within that period, the contract would stand terminated. If, however, oil was discovered in commercial quantities, then too the contract would stand terminated, and a new arrangement was to be concluded between NIOC and the contractor, under which NIOC could agree to sell to the contractor up to 50 per cent of the output at market price minus a discount of 3 to 5 per cent for a period of 15 years. It was estimated that assuming a price of $10 per barrel under the new service contract, the oil company would have to buy the oil at $9.50 per barrel, whereas under the formula contained in the early service contracts, it would have been able to buy the oil at $7.00 per barrel.[126]

(c) *Brazil – model service contract proposed by Petrobras (1976).* In Brazil, after considerable debate, it was decided to invite foreign oil companies to involve themselves in petroleum exploration on the basis of a model service contract, published by the Brazilian national company, Petroleo Brasiliero SA (Petrobras) in 1976.

Under the Petrobras service contract, the contractor was to render technical and financial services. Technical services included the carrying out of all operations required for exploring, appraising and developing petroleum fields, while financial services consisted of provision of funds required for all these operations. The exploration was to be conducted at the risk of the contractor, and the exploration expenses incurred by it could only be recovered if and when commercial production was achieved.

The exploration period was limited to 3 years, and under the relinquishment provisions, 50 per cent of the contract area was to be relinquished at the end of the first year, 50 per cent of the remaining area at the end of the second, and if no commercial production was achieved, the whole area would have to be surrendered at the expiry of the exploration period.

The contractor was subject both to minimum expenditure obligations, and a minimum work programme, particulars of which were to be set out in the appendix to the contract. Exploration programmes and budgets were to be approved by Petrobras.

The contractor was under an obligation to develop any discovery which was regarded as commercial. The test of commerciality was expressed in the following terms:

> 13.5 For the purposes of this contract, a field shall be considered commercial only if its projections of oil production indicate net incomes greater than, or at least equal to the financial responsibilities that this field generates with respect to exploration, development and remuneration re-imbursements. The net incomes of the field under evaluation stem from the gross incomes corresponding to its anticipated oil production, after deduction of all direct costs of production, collection, storage and transportation of oil to the delivery terminal, the applicable overhead and the Brazilian legal severance taxes. The responsibilities correspond to the re-imbursements to meet the exploration and development expenditures and the remuneration due to the contractor, as defined in the present contract. Discount factors to obtain present-day values of incomes and re-imbursements necessary to make sums and comparisons shall be calculated with a reasonable effective annual interest rate.

It was, however, stipulated that the decision as to whether a commercial field had been discovered or not would be made by Petrobras, and such decision would be final. The control and supervision of operations would throughout vest in Petrobras.

The funds advanced by the contractor for exploration would, if commercial production was achieved, be reimbursed without interest, while the funds advanced for development would be reimbursed with interest at an agreed rate. By way of remuneration, a pecuniary fee would be paid to the contractor, calculated on the basis of a formula which took into

169

account the volume of production and the market price of crude oil. Payment would be subject to the "net annual income" and if that was not sufficient, then the liability would be carried forward to the following year.

The contractor would, however, be entitled to purchase at the market price quantities of oil produced from fields discovered and developed by the contractor, limited to the value of the payments which were to be made by Petrobras to the contractor, by way of interest and fee. This right to purchase oil would be subject to suspension in case of a crisis in national supplies of petroleum. The circumstances in which a crisis could be said to have occurred was to be determined by reference to criteria to be applied by Petrobras.

The terms embodied in the Petrobras model service contract were manifestly stringent from the companies' point of view. The provisions to which they took particular exception were the ones that gave Petrobras the final say as to whether a discovery should be regarded as commercially significant or not, and another under which their right to purchase oil could be suspended.[127]

It appears that a number of companies made "non-conforming" bids, and of the forty which had applied, only five groups remained in the field at the time when the first awards were made.

Evaluation of service contracts

The service contract is based on the premise that the host government needs certain services from the oil companies — which it is willing to hire or buy from the companies. The services required are of three kinds: technical, financial and commercial. Governments need the technology of the companies: most also need assistance with financing of operations and marketing of any oil that is produced. Ownership of oil, and of plant, equipment and other assets acquired for petroleum operations, are retained by the government or the national company. As owner, it also retains powers of management. Even the early service contracts, such as the ERAP contracts, provide for control of operations by a joint committee prior to a complete "take-over" of management, which was expected to take place within a period of about 5 years from the date of commercial production.

In an evaluation of the ERAP service contract, while it is conceded that so far as financial returns to the government were concerned it was "absolutely impossible to assert that one or the other type of contract was better for the state", the actual advantages of this type of contract were to be seen at a different level, as it was argued that:

> It is quite clear that the agency contract gives the national company, in other words the country where the operations are performed, a more direct influence over the conditions of oil exploration and exploitation than the concession contract.[128]

While it was further conceded that even under concessions, governments could acquire greater control over operations through equity participation and representation on the board or through the exercise of more extensive regulatory powers, the greatest merit claimed for the ERAP contract was that it led to "a better parallelism in the economic choice of decisions between companies and governments". The importance of this it was rightly argued lay in the fact that "no legal procedure, however well defined, can solve the problem of relations between companies and states if the contract conditions systematically bring about conflict of interest between the parties".[129]

Thus, the early service contracts, embodied a number of mechanisms designed to reduce the possibility of tension or conflict between ERAP and the national oil company. A short exploration period, together with a strict relinquishment programme, reinforced by the provision that any unspent balance of the minimum expenditure commitment would have to be paid over the national company provided safeguards against the contractor's failure to discharge his obligations during the exploration period. The formula for determining whether a well was commercial or not embodied a simple criterion, which left little scope for difference of opinion. If there were to be such a difference, the mechanism which was provided was a variation of a "sole risk" clause, whereby the contractor would drill further appraisal wells on terms to be agreed with the national company. During the production phase, the scope for disagreement was reduced by the requirement that, even before completing "take over" of the management by the national company, all key decisions were to be referred to a "joint operating committee". With regard to marketing, the contractor's obligations were to render "assistance" — elaborate formulae were set out with regard to pricing, so as not to leave it for determination by the contractor. The financial arrangements were so designed as to provide for a phased loan repayment programme from the net cash flow generated by the operations. Thus, the ERAP contract represented an attempt to reduce the conflict arising from decisions being left to the contractor's discretion by making provision for decisions to be taken jointly or by formulating and incorporating "objective" bases or criteria by reference to which decisions would have to be taken.

The advanced service contracts reflect a high level of confidence and capability on the part of the national agencies concerned. They reserve plenary powers of management for the national company, through which it can control all basic decisions. Nor is the reservation of such power "symbolic", since many of the companies concerned have developed sufficiently high levels of capability to exercise relatively effective control. The contracts thus came close to being a contract for supply of services at a fixed price — with the contractor being given no share in the profits nor a risk premium for having assumed the risk of exploration, which even under the advanced service contracts was to be borne by the contractor. An arrangement, however, where no risk premium is provided, is one which only established producers can expect to secure — others, in particular developing countries, intending to enter into service contracts, where the risk of exploration is to be borne by the contractor, would undoubtedly be expected to pay a risk premium — in the form of a share of production or a share of profits, or a right to purchase a portion of the oil at a discount. The size of the premium will depend upon the bargaining skill and strength of the government or national company, and its ability to negotiate financial terms, in the context of alternative discounted cash flow analyses — an aspect of negotiations which is considered in greater detail in the next chapter.

CHOICE OF LEGAL FRAMEWORK FROM AMONG ALTERNATIVES

The development of new forms of agreements presents governments with a range of choice. The effect of this development has been discussed thus:[130]

> The 1960's brought major innovations in the forms of mineral agreements. Most important, the new structures have broken the tight link between ownership, control, and financial risks and benefits that was inherent in the traditional concessions. Arrangements have been negotiated which have re-packaged these elements in ways which were not feasible under the old structures . . . Because ownership and control have become important political symbols in most developing countries, new contractual forms have to be created to allow greater freedom in allocating ownership, control and financial risks in ways that reflect the bargaining power of the parties.

It is important to emphasise that governments in approaching the matter must not only take such steps as they can, some of which were considered in the preceding chapter, to increase their relative bargaining strength, but must clearly identify the specific interests which they wish to safeguard and the mechanisms by which these are to be safeguarded.

"Symbols" are important, not in themselves, but to the extent to which they provide a basis for changes of substance in the relations between governments and companies. In choosing a form a government must look beyond "symbols" to the specific provisions and mechanisms which make up the total "package", and assess how effectively these serve the interests which are to be secured. Companies are seen in the course of negotiations, indeed quite understandably, to demonstrate their "flexibility" by conceding on "symbolic" (or, as they describe it "cosmetic") issues, while holding firm on matters of substance.

As indicated at the beginning of this chapter, a useful check-list for examination of the components of a petroleum development agreement would embrace the following matters:

(a) How long is the duration of the arrangement?
(b) How long a period is stipulated — for exploration?
                                      for exploitation (production)?
(c) How extensive is the area covered by the arrangements?
(d) How are the relinquishment/area reduction provisions structured?
(e) How specific and adequate are (i) the minimum expenditure obligation (ii) the minimum work obligation/minimum drilling requirements during the exploration phase?
(f) How are financial provisions formulated? What is the rate of return for the companies which is assumed by them? How has the rate of return been computed? (This aspect is considered in greater detail in the next chapter.)
(g) Who owns the oil at the well-head?
(h) Who owns the assets required for the purpose of the operations?
(i) In whom are powers of management vested?
(j) What is the criterion by reference to which the question, whether a discovery is commercially significant or not, is determined?
(k) What provisions are made in relation to the development of "marginal fields? Is there a provision enabling one of the parties to undertake development at its sole risk?
(l) How effectively can the government and national company supervise/control operations? To what extent are they involved in the making of key decisions and in the control of each phase of the operations?
(m) What arrangements are made for marketing? How much control does the government or national company have over pricing/marketing?
(n) Is any provision made for supplying petroleum to meet domestic requirements?
(o) What arrangements are made for training and employment of nationals?

173

How effective are these for transferring skills to nationals and for building up national capabilities?

(p) What provisions are made for other benefits to be extended to the national economy?

Every petroleum development arrangement, regardless of its forms, whether it be a licence granted under a statute — often described as a "modern concession", or a joint venture or joint structure agreement, a production-sharing contract, or a service contract, deals with the matters referred to above.

Nearly all of them contain certain basic features in common, which include:

(a) The risk of exploration is borne by the oil company, that is, exploration costs are borne by the company, and are not recovered, unless there is a commercial discovery.

(b) The exploration phase is distinguished from the exploitation or production phase, so that if there is no commercially significant discovery within a stipulated period, the company does not acquire the right to exploit and in fact is obliged to surrender the area concerned.

(c) The company commits itself to a minimum expenditure obligation and in some cases also to a minimum work obligation.

(d) The company assumes the obligation to furnish data and information to the government or the national company.

(e) The company assumes the obligation to develop, contingent on a discovery being appraised as "commercially significant".

(f) The marketing of oil is generally undertaken by the company, subject to provisions regarding pricing.

(g) Provisions are made for division of the financial returns which are generated.

(h) Provision is made for training of nationals and utilisation of nationally produced goods and services rendered by nationals.

The significant differences among the different forms, therefore, are not to be seen in terms only of which form may be financially more advantageous (a question which, as the analysis in the next chapter shows, does not admit of a simple answer), but of the different mechanisms which they embody to secure certain interests and strategic objectives.

Under the concessionary form, traditional or modern, the main interest of the government is seen as the earning of revenue, while control of operations is left largely to the companies, subject only to the exercise of regulatory powers. The introduction of "participation" provisions, such as the carried interest clause, in modern concessions, in fact transforms it into a joint venture or joint structure, and enables the government or the national company to involve itself in management and in the control of operations.

Thus, the distinctive feature of a joint venture or joint structure form is the direct involvement of the government or national company in the management and control of operations. By incorporating sole risk provisions, the government or the national company, reserves the power to develop discoveries, which under a concession could remain undeveloped, if the company should decide not to develop it. Participation in actual operations by government representatives is also an effective mechanism for acquiring technical, and in particular managerial, skills. The "inner workings" of an operation are more likely to reveal themselves to government representatives involved in the operations, than to government inspectors functioning under the regulatory framework.

The production-sharing contract and the service contract formally vest plenary powers of management in the national company. These contracts are so structured as to reduce the possibility of conflict between government or the national company and the foreign oil company since in the taking of key decisions, the national company is suppose to have the final say. The effectiveness of a provision by which plenary powers of management are vested in the national company, depend largely on the instruments which are available for the exercise of the powers. The objective that is important in both these forms is to develop national capabilities so as to undertake operations directly, and for this purpose to acquire expertise, skills and knowledge from the companies through involvement with them in a collaborative relationship under a contractual framework.

While it is a truism to say that "it is absolutely impossible to assert that one or the other of these contracts is better for the state",[131] it is certainly possible to say that a certain set of mechanisms are more effective for securing a set of objectives than another: so that it is possible, given a clear definition of objectives, to make a rational selection from among alternative mechanisms. It is possible, and indeed advisable, to extract certain mechanisms contained in one form, and to incorporate them in another, or to replace a less effective mechanism contained in one by a more effective one drawn from another. The above survey has shown how most of the agreements have been developed and strengthened, through a process of "cross-fertilization", which has led to the emergence of "hybrids" — thus for example, the features of joint ventures have been grafted on to concessions, and to production-sharing contracts, and various provisions designed to ensure rapid and effective exploration under one

175

system, have been readily incorporated in another.

The task of establishing a good legal framework involves not the mechanical imitation of standard precedents, but the imaginative and resourceful putting together in a package of mechanisms which have proved to be the most effective to secure certain general objectives, while safeguarding certain specific interests in each distinct phase of operations.

NOTES

1. *The Exploration for and Exploitation of Crude Oil and Natural Gas in the OECD European Area including the Continental Shelf: Mining and Fiscal Legislation* (OECD Mining and Fiscal Legislation, OECD, 1973); also M. H. Brenscheidt, "Petroleum Legislation in the North Sea Countries", *Texas International Law Journal*, 1976, Vol. II, pp. 281–303.
2. Royal Decree of 8 December 1972 relating to Exploration for and Exploitation of Petroleum in the Sea-bed and Sub-strata of the Norwegian Continental Shelf ("Norwegian Royal Decree").
3. Ibid., Section 31.
4. Interviews with representatives of the Norwegian Ministry of Industry responsible for negotiations with oil companies.
5. The relevant portion of OPEC Resolution XVI. 90 is as follows: "in relation to future contracts their terms and conditions 'shall be open to revision at predetermined intervals as justified by changing circumstances.' Furthermore, such changing circumstances 'should call for the revision of existing concession agreements.' In particular, where provisions for Governmental participation in the ownership of the concession holding company under any of the present petroleum contracts has not been made, the Government may require a reasonable participation on the ground of the principle of changing circumstances."
6. Confidential Memorandum prepared for a Board Meeting of the Texas Co. on 16 August 1946, reproduced in US Senate, *MNC Hearings*, Part 8, p. 105.
7. Middle East Oil (used as Background Paper for 11 September 1950 meeting with oil executives), State Department Policy Paper dated 10 September 1950, reproduced in US Senate, *MNC Hearings Report*, Part 7, p. 131.
8. A. R. Thompson, "Sovereignty and Natural Resources – A Study of Canadian Petroleum Legislation", *Valparaiso University Law Review*, 1967, Vol. I, p. 284 at p. 290.
9. *Petroleum Economist*, April 1976, pp. 130–132 and p. 359 and interviews (July 1977) with representatives of US oil companies operating in Indonesia.
10. NIOC-AGIP Joint Venture Agreement dated 24 August 1957, *Petroleum Legislation, Concession Contracts*, Middle East, Supplement.
11. See the discussion in Roland Brown, "Choice of Law Provisions in Concessions and Related Contracts", *Modern Law Review*, 1976, Vol. 39, pp. 625–642 and Geiger, "The Unilateral Change of Economic Development Agreements," *International and Comparative Law Quarterly*, 1974, Vol. 23, p. 73.
12. Smith and Wells, op. cit., p. 30.
13. This information was obtained in an interview with a representative of the Government of Trinidad and Tobago involved in dealings with the oil companies.
14. The representatives (senior executives) of US oil companies interviewed in July 1977 were unanimous in expressing their opposition to the inclusion of such clauses in petroleum agreements.

15. A. R. Thompson, op. cit., p. 290.
16. Gertrud C. Edwards, "Foreign Petroleum Companies and the State of Venezuela", in Mikesell, op. cit., p. 101 at p. 125.
17. Norwegian Royal Decree, Section 14; New Zealand – Petroleum Act, 1937 (Section 5) and Regulation 6 of the Petroleum Regulations, 1939 (as amended in 1975).
18. The Trinidad and Tobago Petroleum Act, 1969, Section 23; Norwegian Royal Decree, Section 54.
19. Greece, Italy, Sweden, Britain: OECD Mining and Fiscal Legislation, op. cit., pp. 46–47; Petroleum Amendment Act, 1975 (New Zealand), Section 6, Section 13.
20. OECD Mining and Fiscal Legislation, op. cit., pp. 42–45 and pp. 51–54.
21. Norwegian Royal Decree, Section 21.
22. New Zealand – Petroleum Act, 1937 (as amended in 1975), Sections 5 and 6 and Regulation 6 of the Petroleum Regulation, 1939 (as amended in 1975); Trinidad and Tobago, Petroleum Act, 1969 and Petroleum Regulations, 1970, Regulations 44–49.
23. New Zealand – Petroleum Act, 1937 (as amended in 1975) and Petroleum Regulations, 1929 (as amended in 1975), Regulations 22 and 24; Norwegian Royal Decree, Sections 10 and 29; Britain, Petroleum and Submarine Pipelines Act, 1975, Section 26; Norwegian Royal Decree, Sections 38 and 47.
24. T. W. Waelde, "Lifting the Veil from Transnational Mineral Contracts: A Review of Recent Literature", Natural Resources Forum, 1977, Vol. I, p. 277 at p. 279.
25. Australian Petroleum (Submerged Lands) Act, 1967: see A. R. Thompson, "Australian Petroleum Legislation and the Canadian Experience", Melbourne University Law Review, 1967, Vol. 6, p. 370 at pp. 387–388 and "Australia's Off-Shore Petroleum Code", University of British Columbia Law Review, 1970, Vol. 2, p. 1 at pp. 19–20.
26. "Mining the Resources of the Third World: From Concession Agreements to Service Contracts", American Society of International Law, Proceedings of the 67th Annual General Meeting, 1973, Vol. 67, p. 227 at p. 230 (remarks by David N. Smith).
27. Analysis of Recent Mining Agreements, Study Report prepared by the Institut für Auslandisches und Internationalisches Wirtschaftsrecht, Frankfurt for UN, Department of Economic and Social Affairs (obtained through the courtesy of Dean David N. Smith of the Harvard Law School), p. 3.
28. "Mining Resources of the Third World: From Concession Agreements to Service Contracts", op. cit., p. 240 (remarks by Charles J. Lipton).
29. D. N. Smith and L. T. Wells, Jr, "Mineral Agreements in Developing Countries", American Journal of International Law, 1975, Vol. 69, p. 560 at pp. 589–590.
30. Kenneth W. Dam, Oil Resources: Who Gets What How?, p. 14.
31. Robert Fabrikant, Oil Discovery and Technical Change in South-East Asia – Legal Aspects of Production Sharing Contracts in the Indonesian Petroleum Industry, p. 26.
32. Norwegian Royal Decree, Section 17: Petroleum and Submerged Pipe-lines Act 1975, Section 17 and Schedule 4, Clause 14.
33. Sliding-scale area rentals operate as instruments for deterring companies from "sitting on contract areas", that is retaining areas without conducting exploration themselves, since it keeps on raising the cost of such retention. This instrument is used in Norway and also forms part of the 1974 Indian production-sharing contract. An example of fiscal incentives for exploration is provided in the new legislative proposals in Canada, under which the rentals on provisional leases would be refunded for new exploration work, see Statement of Policy, Proposed Petroleum and Natural Gas Act, May 1976, Ministry of Energy, Mines and Resources, Canada.

34. Information obtained in the course of interviews with US oil companies, July 1977.
35. The 1976 Canadian legislative proposals provide for a Progressive Incremental royalty, so that fields would be subject to an increased royalty above a 25 per cent floor rate of return based upon revenues received after deduction of operating costs, and basic royalty and allowances for investment and income tax, with power being reserved to order a reduction where it is considered necessary "to commence or continue production" (i.e. in the case of marginal fields).
36. A. Farmanfarma, "The Oil Agreement between Iran and the International Oil Consortium: The Law Controlling", *Texas Law Review*, 1955–56, Vol. 34, p. 259 at p. 269.
37. *First Report from the Select Committee on Nationalised Industries (Session 1974–75) – Nationalised Industries and the Exploration of North Sea Oil and Gas*, House of Commons, pp. 17–18.
38. Views expressed by US oil company representatives in the course of interviews with them, July 1977.
39. This view was strongly put forward by the agencies concerned with regulating petroleum operations in Norway, in the course of interviews with them in March 1977.
40. Dam, op. cit., p. 14.
41. "There was lengthy discussion on what concessions would satisfy the SAG (Saudi Arabian Government) and how long such satisfaction would continue. Mr Spurlock stated that Saudi Arabs in his experience had always basically favored some part of "partnership" relationship with Aramco and that any agreement would have to take this principle into consideration if it were to last. It was stated that increasing production under fixed royalties only whet Saudi appetites whereas a sharing in the profits or incomes might give the Saudis a feeling of participation which would better withstand the test of time, particularly bad times . . . by partnership Saudis would not demand stock partnership or participation in management but partnership in profits", Extract from *State Department Memoranda of Conversation concerning 1950 (Aramco) Tax Decision*, reproduced in Hearings before the Subcommittee on Multinational Corporations of the Committee on Foreign Relations, 93rd Congress, II sess., on "Multinational Petroleum Companies and Foreign Policy", Report (*MNC Hearings Report*), Part 8, p. 350.
42. M. A. Mughraby, *Permanent Soviereignty over Oil Resources*, is a detailed study of these early joint ventures, see esp. Part II, pp. 45–116.
43. Dam, op. cit., p. 15.
44. C. R. Dechert, *Ente Nazionale Idrocarburi – Profile of a State Corporation*, p. 66.
45. Iran-NIOC-AGIP Joint Venture Agreement, 14 August 1957.
46. Ibid., clause reproduced and discussed in Mughraby, op. cit., p. 92.
47. Egypt-Pan American Agreement, 1963, clause reproduced and discussed in Mughraby, op. cit., pp. 89–91.
48. Mughraby, op. cit., p. 90.
49. Ibid., p. 91.
50. P. J. Stevens, *Joint Ventures in Middle East Oil 1957–1972*, Unpublished thesis – University of London), 1974, Ch. III.
51. Mughraby, op. cit., p. 82.
52. Kuwait-Hispanoil Agreement, 3 May 1967, reproduced in *Selected Documents of the International Petroleum Industry*, OPEC, 1968, p. 155.
53. Dam, op. cit., p. 15.
54. W. G. Friedmann and J. P. Beguin, *Joint International Business Ventures in Developing Countries*, p. 36.
55. Ibid., pp. 38–51.

56. Ibid., p. 41.
57. Ibid., p. 42.
58. Ibid.
59. Ibid., p. 43.
60. Ibid., p. 48.
61. Iran-Joint Structure Agreement with AGIP, Phillips and ONGC, 17 January 1965, Article 3.
62. Iran-Joint Structure Agreement with Amerada Hess, 27 July 1971.
63. Iran-NIOC-Sapphire Joint Structure Agreement, June 1958.
64. Mughraby, op. cit., p. 97.
65. Friedmann and Beguin, op. cit., pp. 49–50.
66. Ibid., p. 44.
67. Saudi Arabia-Petromin-AGIP Contract, 21 December 1967, reproduced in *Selected Documents of the International Petroleum Industry*, OPEC, 1967, p. 190.
68. Ibid., p. 212.
69. Libya-LIPETCO-AQUITAINE/AUXERAP Agreement, 30 April 1968, reproduced in *Selected Documents of the International Petroleum Industry*, OPEC, 1968, p. 195.
70. Dam, op. cit., p. 139.
71. The new rules which came into force in May 1976 in the Netherlands provide that the state would have the right to participate to the extent of 50 per cent in respect of both oil and gas discoveries.
72. *United Kingdom Offshore Oil and Gas Policy*, 11 July 1974, Cmd Paper 5696.
73. Petroleum and Submerged Pipe-lines Act, 1975.
74. M. Adelman, "Is the Oil Shortage Real?", *Foreign Policy*, 1972–75, Vol. 9, p. 84.
75. Questionnaires circulated to a number of Commonwealth governments.
76. Mr Jenkins, MP, in the debate in the House of Commons on the Petroleum and Submarine Pipe-lines Bill, 30 April 1975, Parliamentary Debates, *Official Report*, Fifth Session, 1974–75, Vol. 891, pp. 506–507.
77. David N. Smith and Louis T. Wells, Jr, op. cit., p. 573.
78. David N. Smith, "Mining the Resources of the Third World", op. cit., p. 230.
79. H. S. Zakariya, "New Directions in the Search for and Development of Petroleum resources in the Developing Countries", *Vanderbilt Journal of Transnational Law*, 1976, Vol. 9, p. 545 at p. 552.
80. Fabrikant, op. cit.
81. The problem faced by Venezuela, as the expiry date of the existing concessions approached, of companies "losing interest" and therefore failing to make the necessary investments needed to maintain efficient production, was described by the officials in the oil sector in Venezuela who were interviewed in October 1977.
82. Production-sharing contracts have been extensively adopted in other countries, which include: Bangladesh, Egypt, India, Malaysia and Trinidad and Tobago.
83. The question whether it was the state or the national oil company which owns the oil is discussed in R. Fabrikant, "Production Sharing Contracts in the Indonesian Petroleum Industry", *Harvard International Law Journal*, 1975, Vol. 16, p. 341, thus: "A dispute between Pertamina and the Ministry of Finance as to the nature of Pertamina's petroleum rights has given rise to controversy over title to the oil. Pertamina claims that article 33 means the natural resources belong to the people rather than to the government. Pertamina also claims that the government transferred its petroleum interests in granting a mining authorization to Pertamina. The Ministry's response is that Pertamina has been granted the right to control oil *as the state's custodian* and is, therefore, strictly accountable to the government. These arguments have often been

advanced by rival groups with reference to Pertamina's fiscal responsibility to the Ministry. Recent legislation which imposes special tax rates on Pertamina has resolved this particular issue."

84. Ibid., p. 340.
85. Ibid., p. 341.
86. The 1966 form did not contain an express declaration but the company had "an economic interest": the formulation has been embodied in later forms.
87. Fabrikant, op. cit. (*Harvard International Law Journal*), p. 344.
88. New Zealand – Petroleum Regulations, 1939 (as amended in 1975), Regulations 22 to 24.
89. Standard PSC form, reproduced in Bartlett, Barton *et al.*, *Pertamina-Indonesia's National Oil*, pp. 337–363 at p. 339.
90. Fabrikant, op. cit. (*Harvard International Law Journal*), p. 345.
91. Ibid., p. 317.
92. Ibid.
93. Ibid., p. 336.
94. Ibid., p. 317.
95. Ibid., p. 345.
96. Standard PSC form reproduced in Bartlett, Barton *et al.*, op. cit., p. 352.
97. Fabrikant, op. cit., p. 346.
98. Smith and Wells, op. cit., p. 587.
99. Views expressed by officials in the course of interviews in Norway, March 1977.
100. Fabrikant, op. cit., p. 346.
101. *Petroleum Economist*, 1976, Vol. 43, pp. 120–122.
102. US Internal Revenue Service ruling, reproduced in *Petroleum Legislation: Basic Oil Laws and Concession Contracts in Asia and Australasia*, Supp. 51.
103. Ibid.
104. Information obtained in the course of interviews with US company representatives, July 1977.
105. Ibid.
106. *Petroleum Economist*, 1976, Vol. 45, p. 359, and the Model PSC contract published by Pertamina in 1976.
107. Information obtained in the course of interview.
108. Text of agreement reproduced in *Petroleum Legislation: Basic Laws and Concession Contracts* – Asia and Australasia, Supp. 53.
109. R. Vedavalli, *Private Foreign Investment and Economic Development – A Case Study of Petroleum in India*, p. 195.
110. Information obtained in the course of interview.
111. Dam, op. cit., p. 18.
112. R. Fabrikant, "Pertamina: A Legal and Financial Analysis of a National Oil company in a Developing Country", *Texas International Law Journal*, 1975, Vol. 10, p. 495 at pp. 535–536.
113. H. Zakariya, op. cit., p. 564.
114. Gertrud F. Edwards, "The Frondizi Contracts and Petroleum Self Sufficiency in Argentina", in Mikesell, op. cit., p. 137 at p. 167.
115. Ibid.
116. Bartlett, Barton *et al.*, op. cit., p. 194.
117. Mikesell, op. cit., p. 120.
118. Ibid. and L. Valenilla, *Oil: The Making of a New Economic Order – Venezuelan Oil and OPEC*, p. 129.
119. Report to Venezuelan Congress entitled *Service Contracts before the National Congress*, p. 117.
120. *Selected Documents of the International Petroleum Industry*, OPEC, 1968, p. 381.
121. Ibid., p. 383.

122. Valenilla, op. cit., p. 131.
123. F. Tugwell, *The Politics of Oil in Venezuela*, pp. 107–108.
124. Views expressed by government representatives in the course of interviews in Venezuela, October 1977.
125. J. Montel, "Concession versus Contract", in Mikdashi, Cleland and Seymour (*Continuity and Change*), op. cit., pp. 108–109.
126. J. Amuzegar's report on Iran in G. J. Mangone, *Energy Politics of the World*, p. 297 at p. 323.
127. Information obtained in the course of interviews with US oil company representatives, July 1977.
128. J. Montel, op. cit., p. 114.
129. Ibid., p. 115.
130. Smith and Wells, op. cit., p. 589.
131. Montel, op. cit., p. 114.

CHAPTER V

Financial Provisions of Petroleum Development Arrangements

A central element in any petroleum development arrangement is the package of financial provisions, whereby the financial benefits and risks are allocated between the parties. The financial return to government, referred to in the trade as "the government take", can assume different forms. It may consist of royalties, rents, fees, taxes, or a share in profits or in production. Companies normally determine the profitability of a petroleum development venture by calculating the discounted cash flow (DCF) rate of return. Although the calculation of discounted cash flows is not often resorted to by governments, the financial returns to both parties lend themselves to a discounted cash flow analysis. It becomes important, therefore, for government negotiators to understand this method of assessing financial returns, and more important to understand and assess the validity of the assumptions on which the DCF rate of return is calculated. These assumptions relate to such matters, as probability of discovery, estimated size of discovery, projected levels of production, costs, and future prices.

The DCF analysis deals with financial costs and benefits, as distinct from social costs and benefits. For a government, trade-offs between social benefits and costs on the one hand, and financial benefits and costs on the other, may be significant. Policy-makers and negotiators would, therefore, do well to identify the elements of the social costs and benefits that are involved and in cases where this is possible, to compute social costs in financial terms — in order that trade-offs may be made with due awareness of their financial cost.[1]

While financial provisions may be assessed in terms of the financial returns which they are likely to bring to the parties, it has been perceptively observed that:

> financial instruments should be used not simply as a means for maximising and stabilising revenue, but as *incentives* for meeting the larger package of host government objectives. Governments in developed countries are finding it more and more useful to combine negotiations with financial incentives to achieve certain goals — for

example environmental protection — through a combination of negotiations and incentives such as levies on waste discharge. Surface rentals increasing with time, royalties differential according to the ore content of the ore body involved (or, in the case of petroleum, the estimated size of the reservoir involved), financial penalties and rewards varying with the employment of nationals — all such financial levies and incentives might be used to achieve sectoral objectives.[2]

In order to assess the financial package which forms part of a petroleum development agreement, it is important to grasp the complexities of a discounted cash flow analysis as applied to the particular case involved, and to assess the impact of different types of financial and fiscal devices, not only upon cash flows, but in relation to the different objectives which each of the parties seeks to secure.

I. DISCOUNTED CASH FLOW RATE OF RETURN

Problems of computation

Companies evaluate a petroleum development arrangement in terms of the discounted cash flow (DCF) rate of return which it is expected to yield. This method proceeds on the basis that a project such as one of petroleum development involves substantial capital expenditures before it starts to yield revenue. Further, capital expenditures are expected to continue throughout the period of operations, when in addition the company would have to bear operating costs and be called upon to pay taxes and other government dues. These expenditures and payments would thus represent cash outflows. On the other hand, a point would be reached when cash inflows would commence, not only in terms of profits, but also from depreciation, depletion or amortisation allowances, as may be provided for by the applicable fiscal regime. Thus during the exploration and development period, the annual net cash flow would be negative: thereafter, it could be expected to be positive, that is the cash inflow would exceed the outflow. The difference between inflows and outflows represents the net return. The method of computing the net present value of a venture and of the DCF rate of return is described thus:

> Once a project's cash flow has been calculated it is possible to proceed to calculate its best present value and yield. The basic principle is that £100 receivable today is worth more than £100 receivable in a year's time. This is because £100 invested today will bear interest or profits and so accumulate to more than £100 in a year's time. Alternatively, £100 received today can be used to reduce borrowing thereby avoiding

interest payments as well as reducing debt by £100. To a firm the relevant rate of interest is its marginal cost of finance . . . If a firm's cost of finance is 7 per cent per annum, £107 received in a year's time is worth £100 today, and £100 which will be received in a year's time is worth about £93½ today; similarly, the present value of £100 receivable in two year's time is about £87, and so on. Given the firm's cost of finance, a project's cash flow in any year can be discounted in this way to obtain its present value. The net present value (NPV) of a project is the sum of the present values of the cash flows for all years during the project's life. If the NPV of a project is greater than zero, then the profits are expected to be more valuable than the outlays on the project and to that extent the project is worthwhile.

It will be apparent that the NPV of a project will vary with the firm's required rate of return [cost of finance] which is used to discount the cash flow. In nearly all cases the lower the discount rate the higher the NPV. The yield or rate of return of a project is the rate of interest which if used to discount the cash flow would make the NPV exactly zero.[3]

The DCF rate of return (or the internal rate of return) is thus "that rate which equalizes the present value of the stream of annual net cash outflows with the present value of the stream of net cash inflows".[4] Another formulation defines the DCF profit rate as "that interest rate, which if the money for the original investment were put in the bank instead, would give the investor as many total dollars at the end of the life of the investment as he would have from accumulating the profit cash flow and reinvesting it each year at the same interest rate".[5]

The complexity in determining what is an acceptable rate of return to a company arises from the fact that the company "will usually take into account the probability distribution of each of the variables that determine the expected rate of return".[6]

Among these variables are: probability of discovery, estimated size of discovery, projected levels of production, which may be attained, projected cost and prices, and "political risk". Thus, as regards the probability of discovery, it has been observed:

> In fact, a certain confusion can often be detected in the economic evaluations because this item of exploration costs is overlooked and one only considers the value of the petroleum reserves after they have been discovered. This error would have appeared acceptable in a period when discovery in the Middle East was extremely cheap owing to the huge size of the fields and the very high probability of success. It would now seem that this is no longer possible, either in the Middle East or elsewhere. It is not possible to consider isolated situations, for example, by presenting in respect of a given area only the results of

those companies which have made important discoveries while over-looking all the expenditure incurred by others with no results.[7]

The risks involved in relation to the other variables has been described thus:

> the historically high risks in exploration, technology, politics and economics associated with discovery and production of petroleum are increasing in the more hostile physical environment in which this activity is now carried on . . .
>
> Exploration risk is obvious. Many exploration ventures start but few discover commercially producible reserves . . . Technology risk is not so obvious to the uninitiated. The more easily discovered and produced petroleum reserves are already being developed. Current and future exploration and development is moving to more physically hostile environments with water depths of 2,000 to 3,000 feet, as well as into the Arctic and the stormy and cold North Sea.
>
> New technology is being developed for frontier environments; however, predicting its cost is impossible. Therefore, in frontier areas it is difficult to remain within even the most elastic budgets. The Alaska pipeline cost was estimated at about US $900 million. Substantial delays for political and technical reasons increased the final cost beyond $9,000 million.
>
> Political risk is difficult to quantify. It relates to the stability of governments, the strength of oppositions and their respective policies and ideologies — and to the assessed likelihood of politically motivated change to government-imposed rules applying to industry. Because many petroleum companies operate across national boundaries, they contend with political and economic risks, such as tax law changes both in their home country and the country of operations.
>
> Economic risk arises from usual supply and demand pressures and also from economic aspects of risks already described. Effects of risks are magnified by the long lead times between commencement of exploration, discovery of a commercially producible reserve, and initial production. The worldwide average time lapse between discovery and initial production is seven years with peak production reached after thirteen years. In addition, production from a reservoir is usually continued for twenty years or more.[8]

Companies lay emphasis on risk factors to argue that these introduce uncertainty with regard to their future earnings. They argue with some reason that a lower value should be ascribed to *uncertain* future earnings than to *certain* future earnings. They thus justify their claim for a *risk premium*, that is a higher rate of return for a "riskier" investment. Such risks are taken into account in DCF calculations through the cash flows being discounted on the basis of higher interest rates. The investor in such cases seeks to protect himself against uncertainty either by claiming a higher rate of return throughout the life of the project or a minimum (high) rate, calculated not over the life of the project but over a relatively short period of say, 6 to 8 years. The measurement of "risk" in relation to many of these factors and the determination of the magnitude of the risk premium introduces an inherently subjective or arbitrary element into DCF calculations, for as has been observed "there is no way of estimating the probability distribution of revolutions or drastic changes in government policies".[9]

The negotiations over the magnitude of the risk premium are therefore likely to prove particularly difficult for a government. Its aim must be to introduce some objective elements in an area which is likely to be dominated by the subjective determination made by company representatives. It can seek to do so by collecting information about the risk premium taken into account in other similarly situated countries, difficult though it might be to obtain information of this kind. It may call upon the company to spell out the risks which it is taking into account in its calculations. Some of the risks can be reduced, or eliminated, by certain commitments being made by the government, or the adoption of certain mechanisms. Thus, it could be contended that the risk premium should be reduced where a quicker pay-out period is provided for, or where there is a provision for international arbitration for settlement of disputes.

There may be cases where,

> If the history of a country is such that companies have no confidence in the contractual agreements made by the government, the company may not be willing to make an investment or to carry out exploration activities even if the potential returns on an economic basis are very large.[10]

Equally, there may be cases where a country might not be willing to pay as high a risk premium as is demanded. It is in such cases that government may have to consider the option of undertaking exploration under arrangements in which companies would play little, or no, part.

Apart from the difficulty presented by the calculation of risk premia, assumptions relating to other variables on the basis of which DCF calculations are made need to be carefully assessed. Since these assumptions relate to projections about costs, prices and quantities to be produced in the future, there can be considerable divergence in assessments which would be reflected in the earnings and production profiles that may be drawn up. Thus, it is observed that

> there will usually be a wide range of cash flows which result from a particular project since there are likely to be several different estimates of factors determining the cash flow, such as the initial

capital costs, future demand for the project, the actions of competitors, the operating costs over the life of the project.[11]

Government negotiators must, therefore, analyse the DCF calculations made by the companies and so make their own independent assessment about the validity (or plausibility) of the assumptions on which the calculations are made. It is, therefore, urged that

> . . . projected rates of return and the technical and economic estimates on which they are based, can and should be checked by geologists, mining engineers and economists working for the governments of the host country and any differences in their evaluation of the project reconciled.[12]

Where expertise to make such assessments is not available within the government, there is a strong case for seeking the assistance of independent consultants.

Comparisons of different arrangements on basis of DCF rates of return

To the question which type of petroleum development arrangement is most advantageous for the government, it is difficult to provide a simple answer on the basis of a comparison of the different types of arrangements,

> for although such comparisons may provide useful information, an examination of existing contracts reveals a bewildering combination of terms covering tax and royalty rates, tax holidays, depreciation schedules, loan-equity ratios, government and local participation . . .[13]

A useful comparison can, however, be made of different types of arrangements, on the basis of the DCF rates of return which these offer to the companies. This would provide governments with some indication of how a company is likely to respond in the course of negotiations to different proposals, for

> there is ordinarily a range of expected rates of return within which bargaining can take place. During the bargaining process, of course, no company is going to reveal the price below which it will never make an investment and surrender an exploration concession on which it may have already spent millions of dollars. And if the company bows out, it may not be easy for the host government to find another prospective investor who would need to spend a substantial amount of money before entering into serious negotiations. It is, therefore, important for the host government to have knowledge of the limits within which successful bargaining can take place.[14]

A company's decision to undertake a petroleum development venture is said to depend primarily on three considerations:

First, what is the expected (DCF) profits rate on such investment? Second, what is the expected profit rate on its alternative exploration and development investment possibilities? Third, what is the minimum expenditure profit rate it requires for expansion and development expenditures?[15]

It becomes important, therefore, to compute what the going DCF rates of return are likely to be on different types of arrangements which are currently being entered into. If these computations are to be useful for the purpose of assessing responses, it would be important to base them on assumptions similar to those on the basis of which the companies have made their calculations.

A number of difficulties involved in making these comparisons should, however, be noted. First, the same assumptions as to cost cannot be applied to different areas. Thus, for example, while the cost of developing a new oilfield in the North Sea is estimated to be in the range of $6000 to $7000 per daily barrel, that in the case of Indonesia or Egypt is estimated to be in the range of $2000 to $4000 per daily barrel. The variation in costs is attributable to differences in physical conditions. Among the elements which affect costs are field depths, water depths, distance from shore, weather conditions, magnitude of reservoir and production potential of the reservoir.

Secondly, the impact on cash flows of the fiscal regime not only of the host regime, but of the home country of the oil company concerned needs to be assessed. As will be evident from a consideration of the different types of fiscal levies, these can have a significant impact on cash flows accruing to the company. Among the host government's fiscal provisions, those that have a particular relevance are those that effect cost recovery; provisions, for example, relating to depreciation, amortisation and depletion.

Depreciation being a deductible cost, the allowance made in this respect reduces the cash outflow to that extent. Thus, by allowing high depreciation in the earlier years, or allowing companies to write off an asset before the expiry of their presumed economic life, the tax burden can be eased on companies in the earlier years. A device, for example, permitted in the post-1976 Indonesian contracts, which while making the overall terms more onerous for companies, allowed capital costs to be depreciated over an average of 7 years using the double-declining balance method.[16]

Amortization, in the context of petroleum agreements, refers to the write off of capitalised expenditures that do not represent tangible assets subject to depreciation and is another item in respect of which deductions are provided for. The more recent agreements, such as the 1974 Egyptian-

Esso agreement analysed in the previous chapter, specifically excluded such items as interest on investment, financial fees and bank charges.

A depletion allowance is a deduction provided for under certain fiscal regimes. Depletion in the sense of the cost of bringing the deposit into production is no more than amortisation. A form of depletion allowance provided for in the United States, termed "percentage depletion" enables the company to deduct a fixed percentage of the gross or net income for this purpose.

Of direct relevance to the calculation of cash flows are home country fiscal provisions which allow a foreign tax credit, a depletion allowance or expensing of intangible drilling costs. Generally, inadequate attention is paid by government negotiators to the home country fiscal regimes, principally because of inadequate knowledge or information about them. Companies can derive substantial benefits by offsets against the tax payable in the home country of tax payments made by them overseas. Thus, for example, US companies where they can prove that they have an economic interest in the petroleum involved, become entitled to a depletion allowance of the order of over 20 per cent of the gross income of the company in the United States. When they can prove direct interest in a part of the production they can deduct "intangible drilling expenses".

In the case of a Canadian company, it is suggested that the impact of Canadian Income Tax Act will depend upon the answer to the following questions:

(i) Is the concession a foreign resource property within the meaning of the Act?
(ii) Will expenses incurred thereunder constitute foreign exploration and development expenses under the Act?
(iii) Will income thereunder be foreign resources income as described in Section 66(4)(b)(ii) of the Act?
(iv) Will amounts paid to the host country be considered taxes with respect to which a foreign tax credit will be available under Section 126 of the Act?[17]

The significance of these questions, for Canadian tax purposes, is explained thus:

(i) If the concession is a foreign resource property, acquisition costs will be deductible as foreign exploitation and development expenses, and proceeds received upon disposition of the concession will be taxed as income. If the concession is not a foreign resource property, acquisition costs may be considered a non-deductible capital expense and, similarly, proceeds of disposition may be treated on a capital gains basis.
(ii) If expenses incurred in operations under the concession are foreign

189

exploration and development expenses, they will be deductible according to the rules set forth in Section 66(4) of the Act, which may not be as beneficial to the taxpayer as if they were treated as ordinary expenses incurred in producing income.

(iii) Foreign exploration and development expenses are deductible to the extent that the taxpayer has foreign resource income as described in Section 66(4)(b)(ii) of the Act. It is accordingly important that income from the concession be considered as income within the meaning of that section, if the taxpayer will have foreign exploration and development expenses.

(iv) If income taxes are levied in the host country at a rate not less than the Canadian rate, and if credit therefore is allowed under Section 126 of the Act, income from the concession may be brought to Canada free of Canadian income tax. If credit is not given for all payments to the host government, additional Canadian taxes will be imposed, and the project will be less attractive.[18]

Other provisions which can affect cash flows are those relating to state participation, and those which subject the company to the obligation to comply with directives of the government as to the rate of depletion, or to meet domestic requirements of petroleum at a deflated formula price. Thus, according to an expert evaluation, while the post-1976 Indonesian terms are no more onerous than those in force in the North Sea, the aspect which companies reacted to particularly adversely in Indonesia was the requirement to supply crude, about 20 per cent of the total quantity produced, to the domestic market at a nominal price of cost plus 20 cents per barrel.

In Appendix A to this chapter is set out a comparative table giving DCF rate of return calculations relating to certain selected countries. The assumptions on the basis of which the calculations have been made are stated in the Appendix.

Concept of "fair rate of return" and mechanisms designed to tax excess profits

The experience of the established oil-producing countries shows that one of the principal sources of conflict between governments and oil companies has related to sharing of the "rent" or "surplus", that is the sums which are left over above the current expenditures necessary to produce output.

It is inherent in the nature of a petroleum development venture, as of most mining projects, given their relatively long duration (currently ranging from 15 to 25 years), that the original commercial assumptions on which the financial provisions were based between the government and the company cease to be valid under the impact of changing circumstances, often resulting in significantly greater returns to the company than might have been anticipated.

This can occur in a number of different situations. At the time when the company acquires the concession or enters into the agreement, the probability of discovery and the size of the expected discovery that may have been taken into account, may have been much lower than might actually turn out to be the case. The risk premium may thus appear in retrospect to have been too high. Viewed from a different angle, the supply price which is charged initially by a company for its capital, its technical and managerial expertise, and its "risk-bearing", which tends to be higher, the greater the perception of risk in relation to the expectation of gains, may after discovery, especially if large discoveries are made, appear to the government to be *too high*. Also, in the post-discovery situation, other sources of supply may become available at a lower price. The current supply price, post-discovery, therefore, does tend to fall. The company, which is party to the original agreement, thus finds itself in a position where it is seen to be earning a substantially higher return than was originally anticipated, or in any event, higher than the current supply price for the factors which are provided by it.

Then again a situation might arise where due to changes in market conditions, there is an escalation in price of the oil significantly higher than that which was anticipated when the agreement was entered into.

With regard to the first type of situation, it has been argued that

> there is no reason . . . why the oil companies should earn greater profits than is necessary to continue to induce them to continue investing in the country and to compensate them for their initial risks. In other words, if excess profits are to be earned, a producing country can complain of exploitation to the extent to which it is unable to obtain them for itself because of the superior bargaining power of the oil company: if it could obtain them, the use of its resources would thereby become more profitable to the country.[19]

The companies on the other hand question the relevance of the concept of "excess profits" in relation to petroleum operations, arguing that the apparently high returns on a discovered field does not take into account the losses borne by them on their abortive exploration efforts, and that the apparently high return represents the risk premium or reward earned by them for having borne the risk of exploration.[20] There is indeed in such a situation, no objective yardstick by which what is a "fair" risk premium can be determined, for it has been observed that

> . . . the rate of return that is derived from resource operations, and which is necessary to induce companies to continue investing in a

country once petroleum has been found in large quantities, differs substantially from that necessary to induce and undertake high-risk investments. There is simply no way to measure the proper compensation for risks under these circumstances.[21]

The issue is thus ultimately left to be determined by bargaining. With regard to the second type of situation, namely the generation of windfall profits due to a substantial rise in price, recent experience has shown that companies, however grudgingly, have begun to accept the position that in such situations where the "surplus" generated is substantially larger than was anticipated, they would not be justified in retaining all or the bulk of such "windfall" or "excess" profits. While company representatives still argue that *in principle* there is no reason why they should not be regarded as *entitled* to retain such windfalls, they admit that in practice it is politically intolerable for governments to allow such "windfalls" to be appropriated by the company. Governments have argued that such higher than anticipated returns should accrue to the state which is the owner. The principles of "equity" and permanent sovereignty over natural resources are invoked in support of the state's claim.

The companies were prevailed upon to concede to the government view by the OPEC countries. The subsequent efforts of the new oil-producing countries, Britain and Norway, to adopt measures to appropriate the "windfall", which according to the governments had been created by the OPEC-led increase in the price of oil, were initially met by concerted opposition by the companies.[22] The mechanisms, however, which were adopted by legislation in 1975 in Norway and Britain, merit examination, as they can be seen as having gone some way in establishing the concept that when the level of profits generated are substantially higher than an anticipated "fair rate of return", the whole, or in any event, the greater part, of the "surplus" or "excess profits" should accrue to the state, and not to the company.

The rationale for the Norwegian Special Tax on petroleum was stated by the Finance Committee of the Norwegian *Storting*, thus:

> The unusually sharp rise in petroleum prices, particularly since the autumn of 1973, has created an entirely new profit situation, also for those companies producing petroleum in the North Sea. Even though the costs for oil and gas production have also risen sharply, production is far more profitable than it was on the basis of former crude oil prices. This entirely new profit situation is the basis for the Government's view that the taxation of oil companies should be altered so that a greater share of this additional revenue will accrue to the Norwegian society than is the case according to existing rules.
>   *The majority* points out that the petroleum resources on the

Norwegian continental Shelf are the collective property of the Norwegian people. In the view of *the majority*, Norway's desire to secure a substantial share of the values created on the Shelf is therefore not unreasonable.[23]

The case for the British Petroleum Income Tax was presented in Parliament in the following terms:

> . . . the original licencees are subsidiaries of foreign owned companies, and there could be a large balance of payments loss through remittance overseas of their share of excess profits unless corrective action was taken . . . In fact the only way of doing this is by increasing the Government's share of the profits.[24]

The Norwegian bill relating to taxation of submarine petroleum resources (Parliamentary Bill No. 26) provided for the introduction of a Special Tax, aimed at "windfall profits" of 25 per cent of income remaining after payment of royalty (10.5 to 12 per cent under the "old licences", 8 to 16 per cent under the new licences), and ordinary corporation tax of 50.8 per cent. A deduction would be permissible of a special investment allowance equal to 10 per cent of the purchase value of all installations and equipment taken into use over the preceding 15 years. In addition, 100 per cent depreciation would be allowed of production and transport installation costs over a period of 6 years and deduction of past losses (for up to 15 years) with a limit in any one year of one-third of losses carried forward in the previous 14 years. The "ring fence" concept was adopted to prevent losses being made elsewhere in the world being offset against Norwegian tax liability, so that losses from off-shore activities within Norway could be set off against profits, as well as up to half of the losses suffered from on-shore activities in Norway. The 1975 legislation also introduced the concept of "norm price" to be fixed every 3 months by the Petroleum Price Board as an average of the "arm's length" prices of crude oil in the free market.[25] The net earning on which taxes could be based would be calculated on the basis of the "norm price".

The British Oil Taxation Act, 1975, provided for a Petroleum Revenue Tax (PRT) to be levied at a single rate of 45 per cent as a prior charge on gross profits from production, before the 52 per cent corporation tax. The tax would be applied on a field-by-field basis. As an incentive for development of marginal fields, provision was made for waiver in respect of any field, where, in any year, it would reduce the return on capital expenditure, measured on the basis of the historical cost, before corporation tax to less than 30 per cent. Above the 30 per cent cut-off point, there is a "tapering provision" which would in effect assure that PRT will not be greater than 80 per cent of the amount by which the profit exceeds 30 per cent of the capital expenditure to date. It is estimated that

> the 30 per cent safeguard will have the effect, taking into account the deductibility of part for corporation-tax purposes, that PRT will not cause the rate of return on capital invested to fall below about 15 per cent.[26]

It is noteworthy that the rate of return to be taken into account for assessing PRT is based on capital expenditure (measured in terms of historical cost) and not a DCF rate of return.

Under the PRT system, deductions were provided for in the form of an oil allowance of 1 million tons a year, subject to a maximum cumulative total of 10 months' taxes. The allowance, representing the money value of the oil at the market price, would constitute a deduction. The provision for the oil allowance is explained thus:

> The oil allowance is explicitly intended to maintain the economic viability of marginal fields, though it has a somewhat broader reach. All fields will benefit from the allowance, but small fields may benefit proportionally more, because however large a field may be, the allowance is limited to one million tons a year. Though water depths, distance from shore, and other factors bear on whether a field is marginal, the size of the field is perhaps the most important factor.[27]

In addition, further provisions were included in the Oil Taxation Act to cushion the effect of PRT and to reduce its disincentive effect, in particular in relation to marginal fields. Thus, an "uplift" of 75 per cent of the capital expenditures was to be allowed. This device enables the company to achieve a quicker pay-out, a feature to which companies attach considerable value. Normally all capital expenditures are not deductible in the first chargeable period of 6 months (and further deductions are to be carried over to subsequent tax periods). An "uplift" is an additional percentage which can be deducted in the first chargeable period. An "uplift", therefore, allows a quicker pay out and postpones the onset of tax liability. The impact of the uplift provision on profitability calculations of the companies is assessed thus:

> . . . the 75 per cent uplift provision will significantly cushion any disincentive effect of PRT. Although uplift is unlikely to favor marginal fields more than other fields (except to the minor extent that capital expenditures are larger relative to operating income in marginal fields than in other fields), nevertheless it is the marginal fields that have to compete with investments elsewhere in the world and therefore any reduction in effective tax rates that applies to all fields will encourage development of marginal fields.[28]

194

A comparative analysis of the British and Norwegian systems, based on four hypothetical fields of different sizes and with different costs, has led to certain preliminary and tentative judgments being expressed in the following terms:

> At current oil prices and tax rates, the two systems will probably yield similar percentage government "takes" (taxes and royalties) and similar company discounted cash flow (DCF) rates of return. The UK system gives a higher take for all fields, but if the smallest field (which pays no PRT) is excluded, there is less than 2 percentage points difference in each case. Company returns tend to be lower under the UK system, except for the 200 million barrel field . . . Both systems lead to somewhat lower percentage takes from the smaller fields than from the larger fields. The UK government's total tax and royalty take is very insensitive to changes in the rate of PRT whereas Norwegian government take is readily varied by altering the ST rate . . . These differences occur essentially because ST operates as a tax additional to royalty or corporation tax, whereas the more complicated PRT is expressed against UK corporation tax and is subject to various safeguards and exemptions . . . (in terms of sensitivities to price changes), the general effect is for return from the larger fields to be marginally lower and take to be marginally higher under the UK system . . . however, the two tax systems lead to roughly equal sensitivities of company returns to price changes . . . [all the above conclusions relate to equity financed projects] . . . Under a debt-financing regime substantial differences in government take and in profitability could appear because the PRT legislation does not allow interest payments and other financing charges to be offset against profits as they are in the Norwegian system [under PRT such charges are supposed to be allowed for in the "uplift" provisions of capital expenditure]. Consequently, the higher the proportion of debt-finance the lower the government take seems likely to be under the Norwegian system as compared with the British.[29]

It is noteworthy that while company representatives expressed a strong negative response when taxation proposals were originally introduced in 1974 to enable governments to appropriate a substantial part of "windfall profits", they have subsequently admitted that the PRT (specially with its safeguards and deductions) represents a fair and workable fiscal regime. They remained unwilling to admit, however, that a contractual provision could be incorporated in petroleum agreements (in particular in production-sharing contracts) which could operate in a manner similar to the PRT, or that such a provision could contribute towards stabilising contractual relationships between oil companies and governments.[30]

The approach whereby at least a substantial part of the surplus over and above a fair return to the producer should accrue to the state, as the owner of the resource was adopted by the British Columbia Energy Commission and is implicit in the Progressive Incremental Royalty system introduced by the new Canadian Petroleum and Natural Gas Act in 1976.

The British Columbia Energy Commission was requested in 1973 to give its advice on the following matter:

> present and anticipated royalties payable to the Province as the owner of British Columbia natural gas and whether the said royalties fully reflect the value of such gas after due allowance for costs and risks incurred in its discovery and removal.[31]

The response of the Commission to this question was expressed in the following terms:

> The owner of a resource obviously can only be properly compensated for it if the resource is fairly priced on the market and the surplus over and above a fair return to the producer is paid to the owner; it is the function of a royalty system to ensure that the owner receives this surplus.[32]

The Commission received evidence of producers setting out DCF calculations. Upon examining the evidence produced by the companies and other witnesses, the Commission applying the above principle, in a situation where an increase of 8 cents per m.c.f. in this price had been allowed, arrived at the following finding:

> While analysis suggests that some of this increase should go to the producers who are not receiving an adequate return at the present prices, it is clear that the provisional royalty should be revised so that the surplus generated by these windfall price increases will accrue to the Province.[33]

The Commission accepted that producers were entitled to a "fair" or "adequate" return, the principal criterion for "fairness" or "adequacy" being that this should sustain the desired rate of exploration and development. The companies had sought to adduce evidence that the existing rate of return was not adequate to sustain the level of exploration and development required to meet the province's needs. The Commission recognised the difficulty inherent in determining what the "fair" rate of return and the optimal rate of royalty should be, and that this depended on changing circumstances. They, therefore, proposed that these questions be made subject to annual review by the Commission, thus:

> Industry cannot provide empirical data establishing a relationship between price and discovery because such data does not exist. It is, therefore, our view that this Commission should maintain a constant review of prices, exploration effort and discovery rates so that there

will be an informed basis for determining future field prices and royalty rates in the course of [an] annual review . . .[34]

When the next occasion for review arose in 1975, the Commission received evidence on rates of return and on the "netback" (return per unit of production after operating costs and income taxes) which companies earned at existing price levels, and upon consideration of the evidence, recommended in its report submitted in 1976 that

> higher prices for old and new gas and increased netbacks to the industry on crude oil, together with a mechanism for revising prices and netbacks in the future, are necessary incentives to provide higher levels of oil and gas production and are justified in the public interest.[35]

The British Columbia experience indicates that the recognition of the relevance of the concept of a "fair" or "adequate" rate of return can, in certain cases, provide support to a company's claim for a higher price or greater return, while in other cases it can be relied upon by the government to tax "excess profits". The Report noted the difficulty presented by substantial variation in the evidence submitted by companies, so that the validity of the underlying assumptions were subject to question under cross-examination. Thus, "operating cost, development cost and in particular finding cost assumptions were questioned", so that;

> With regard to finding costs, there was considerable questioning because the CPA (Canadian Petroleum Association) and Shell estimates, both based on probable reserve estimates credited back to the year of discovery, were much higher than other estimates . . . presented to the Commission in this and past hearings by many of the active explorers in British Columbia. To some extent the CPA estimates overstate finding costs because their analyses do not take into account any future appreciation of resources which may be credited back to the years they examined. In the Shell analysis a different crediting back methodology is used, but this has results in high unit finding costs. However, other explorers who together account for a large share of activity in the province, perceive lower finding costs than the Shell or CPA industry average figures support.[36]

The comparison of DCF calculations presented by different companies thus can be useful in giving a government a rough basis for assessing the validity of the assumptions on which the calculations are based. Assuming that the calculations have been made independently by each of the companies, the absence of large variations would *prima facie* indicate a degree of objectivity, whereas wide variations should call for probing at greater depth the validity of the assumptions on which the calculations are made.

The approach whereby a substantial part of the profits above certain

197

levels should be appropriated by the state is also implicit in the Progressive Incremental Royalty (PIR) mechanism introduced by the new federal Petroleum and Natural Gas legislation (applicable to federal lands) in Canada in 1976. The Statement of Policy made in support of this legislation presents the rationale for the PIR (a levy additional to the 10 per cent production royalty and national income taxes) in the following terms:

> If one considers oil and gas reserves as a capital asset which the nation contributes to the industry, then the 10% royalty might be regarded as a safeguard minimum income. PIR ensures that the Government return will be greater on deposits which are unusually rich ... the rate of PIR (increases) with the rate of "profit". No PIR is imposed on annual profits of less than 25%, because this is the minimum rate which is considered sufficient to yield a reasonable return on both development and exploration expenditures. For profits higher than this the PIR rate rises quite sharply, but levels out at rates approaching 40%.[37]

Among developing countries, Papua-New Guinea in its statement of mineral policy affirmed its adherence to the following principles:

> one, that the foreign investor in resource projects should be allowed a reasonable return on investment, and two, that "the lion's share" of profits above that level should belong to the people of the country.[38]

Application of these principles is found in recent mineral agreements to which Papua-New Guinea has been a party, as well as in the terms and conditions on which it has granted rights for petroleum exploration and development. Thus, the re-negotiated agreement of 1974 between the Government of Papua-New Guinea and Bougainville Copper Limited, and the agreement of 1975 with Dampier Mining Company Limited relating to the Ok Tedi deposit (both relating to copper and gold deposits) incorporated a provision for an Additional Profits Tax. In the Bougainville case, the additional profits tax becomes chargeable whenever profits net of company income tax exceed a rate of return exceeding 15 per cent of capital employed; in the Dampier ("Ok Tedi") agreement, the company is required, after the investment recovery period, to pay a marginal rate of 70 per cent tax (comprising the normal company income tax plus an additional profits tax) on income accruing in any one year in excess of such income as would give a 20 per cent internal rate of return on the project.[39] The new petroleum licensing terms introduced in 1977 in Papua-New Guinea pioneers an Additional Profits Tax which becomes payable when returns on investment exceed 20 per cent money return on total funds invested. The financial provisions assure a quick pay-out and are otherwise so structured as to make the overall financial package acceptable to oil companies, and agreements embodying these terms have reportedly been concluded with Exxon and Philips.[40] Appendix B sets out the principal elements of this fiscal regime.

It has been noted in the preceding chapter that a basic weakness of the standard PSC is that it does not contain any mechanism to deal with the problem of windfall profits, which may arise due to a sudden price escalation, or due to a substantially higher rate of return being generated than was estimated when the agreement was entered into. It is, therefore, advisable for some mechanism to be incorporated in these agreements to deal with this situation. This can be done by making the agreement itself subject to legislation which provides for payment of a tax in the nature of an excess profits tax, as has been done by the Malaysian PSCs of 1976, or by incorporating a contractual provision, whereby the company would undertake to make special payments, representing the greater part of any surplus over the profits required to provide an internal rate of return, of say 15 to 20 per cent, which may be stipulated in the contract.

"Adequate rate of return" to sustain desired level of exploration and development

Governments are concerned to maximise government take from petroleum operations. Since in most cases, however, oil companies are expected to invest in exploration and development, the government while seeking to maximise its revenues, must provide for an "adequate rate of return" to the companies in order to induce them to invest in petroleum exploration and development. As has been observed in the case of mining ventures,

> a mining company contemplating a major investment is likely to be most influenced in its decision by whether the expected rate of return on investment is likely to exceed a certain minimum required rate.[41]

What the required rate may be in respect of a particular region is a matter which admits of varying assessments, as is evident from the following review:

> At one extreme, a study by the Aerospace Corporation states that "historically" oil companies have had to earn approximately a 10% pre-tax rate of return on overall investments in order to attract debt and equity capital . . . This rate is calculated assuming no inflation and thus the actual value in the future may have to be higher to compensate investors for "inflation" . . . Another reported estimate is that a 14% rate of profit (presumably after taxes) would be sufficient to encourage oil and gas exploration in the United States. At the other extreme, a petroleum trade journal claims that for

exploration and development in the North Sea, the companies would require a 25% DCF payment rate.[42]

A range of estimates made for the North Sea, assuming development costs to be $8000 per daily barrel, indicates the company's DCF rate of return as extending from 13 per cent in Norway to 20 per cent in the United Kingdom. Similar studies made for Egypt and Indonesia, assuming development costs of $2000-$4000 per daily barrel, indicate DCF rates of return to be in the 15-20 per cent range for Egypt and in the 17.28 per cent range for Indonesia. In the judgment of a consultant advising the State Legislature of Alaska, "DCF profit rates in the neighbourhood of 20% and up on exploration and development of Prudhoe Bay oil should be more than adequate to meet the companies' requirements".[43]

The required rate of return expected by companies will, therefore, vary from region to region, and is understandably likely to be higher in the case of those developing countries, where there have been no (or only a few) earlier discoveries. If, however, companies were to expect rates of return beyond say 25-30 per cent, the question may well be asked in the terms in which it was posed to the Alaska State Legislature:

> If, however, a DCF profit rate of 20% on exploration and development together would not be adequate to induce major oil companies to invest in Alaska, this should raise a serious question. Namely, whether from the point of view of the people of Alaska simply leasing prospective oil lands to the companies is the best method of developing the state's oil resources? After all, a 20% per annum DCF profit rate ultimately implies that for each $ initially invested by a company, its total profit at the end of a 25 year period would be $95. Since the State itself clearly has a cost of capital which is far below 20% per year, it may well be that if the State has to pay 20% per year or more to get exploration and development carried out by the companies, it would be better off considering alternative methods of exploiting its oil resources.[44]

II. GOVERNMENT TAKE: ALTERNATIVE FISCAL AND FINANCIAL MECHANISMS

While the overall assessment of a financial package (made up of financial and fiscal provisions) may be made in terms of the magnitude of the government take or of the DCF rate of return which is likely to accrue to the company, the elements which make up the package need to be separately assessed in terms of their impact on particular objectives and interests which each of the parties wishes to safeguard.

Thus, for example, a government would favour financial and fiscal devices, which would ensure accelerated exploration, while a company would prefer provisions which left it maximum freedom of action. A company is in general likely to be more amenable to accede to the government's proposals for higher revenues after discovery, than to proposals for higher exploration expenditures and more onerous work programmes. Similarly, while governments may favour greater control over the disposal of the crude oil and over marketing, or ensuring supplies to meet domestic requirements, companies "are apparently prepared to give up a number of percentage points in tax or royalty levels (or in the price they pay the government for its share of crude oil) in order to be certain that they will have the resulting crude oil to feed into their downstream operations".[45] Governments may wish to ensure certain minimum payments by way of royalty to accrue to them, regardless of profits, while companies, having regard to the fact that a royalty formula which operates as a fixed charge upon costs may seriously affect profits when prices are low may well prefer a flexible charge related to profits even if this resulted in a higher expected overall royalty payment. Companies prefer a quicker pay-out and thus attach considerable value to accelerated depreciation and other devices which give it a higher initial rate of return, to the extent that they are disposed to agree to compensate by paying higher taxes (related to profits) in later periods. In some cases, however, "the host government may have an urgent need for income in the short run; this could be balanced against a somewhat higher return for the foreign investor in the long run".[46]

The form in which the government receives its take can also have an impact on companies' financial returns through the effect that it has on the companies' tax obligations in its home country. Thus, a US company, in order to deduct intangible drilling costs for oil and gas, or to get the benefit of a depletion allowance, must establish that it has an "interest" in the oil or gas.[47] This was one of the conditions that led US oil companies to prefer the "joint structure" or "contractual joint venture" or a production-sharing contract, under both of which they owned a part of the oil, to a service contract under which they did not. Similarly, the taxes which were paid on their behalf under production-sharing contracts entitled them to tax credits, until the US Internal Revenue Service ruled that such payments were in the nature of royalty and did not qualify for a tax credit.[48] By the US Tax Reduction Act it appears that a number of tax advantages enjoyed by the US oil companies in relation to foreign expenditure were curtailed, thus:

percentage depletion was repealed for the major oil companies; limits were imposed on the amount which smaller companies could claim and this depletion was gradually phased down to fifteen per cent . . . other charges were adopted that will prevent intangible drilling expenses incurred abroad from being offset against domestic tax for tax purposes . . . finally, the value of the foreign tax credit was reduced.[49]

It is, therefore, important for government negotiators to be aware of the home country tax regime applicable to the companies it is dealing with, so that it may be able to assess the true value to the company of a financial provision. In some cases, in particular those where a foreign tax credit is available, the host government should appreciate that a reduction in tax rate would simply mean that the US Treasury collects what the host country forgoes, or conversely that a higher rate of tax does not increase the financial burden on the company, as it can obtain a tax credit in its home country and, thus, offset these payments against its tax liability in the home country. It is, therefore, advisable for governments, if need be, to obtain independent professional assessment of the impact of home country tax provisions on the company concerned.

It is also important when considering alternative fiscal and financial mechanisms to keep in view certain basic considerations. The conceptual framework within which financial provisions were designed has itself undergone basic change. The basic premise on which the pre-Second World War concessions, which provided only for a fixed royalty per ton to be paid to the state, was that the royalty constituted the consideration for transfer of ownership of the petroleum to the company, which then was entitled to dispose of it and earn the entire profits which accrued from such disposal. This premise itself was challenged when new arrangements were proposed. Thus, in the fifties, the state asserted that in addition to a royalty, it should also be entitled to a share in the profits. In conceding this, the companies in fact accepted a new basic premise, namely that the state, despite its formal transfer of ownership of the oil in consideration for a royalty, retained the right to appropriate a share of the rent/surplus profits generated by its disposal. The developments of the sixties and in particular of the seventies bear evidence to a more radical basic proposition being successfully asserted by governments; namely, that a company, investing its capital and skills, should be entitled to a fair rate of return on its investment, and that any surplus over the designated rate of return should accrue to the state as the owner of the resource. While it may be conceded that "it is not sensible to attempt to siphon off for Government (or any other owner of the mineral rights) *all* the excess revenue above the designated rate of return", for the reason, among others, that "to do so is to sap the incentive of the mine operators to mine efficiently",[50] the basic criteria by which a financial package may be judged is whether they ensure the maximum financial returns to the government, consistent with allowing a return to the companies, *sufficient* to attract the companies to undertake and efficiently conduct the desired level of exploration and development.

It is in this background that the effectiveness of alternative fiscal and financial mechanisms should be assessed. The principal mechanisms from among which a choice has to be made are: royalties, taxation, and participation. Other mechanisms which are found in petroleum development arrangements and which merit examination are: bonuses, rentals, mandatory surrender provisions, provisions requiring domestic procurement of goods and services and supply of petroleum to meet domestic needs, as well as provisions requiring financial contributions to be made towards training of nationals and developing of national capabilities in the petroleum field.

Royalties

A royalty, in its original conception, was the delivery in kind of a specified part of the resource extracted to the ultimate owner of mineral rights, which in most cases was the state. In the case of petroleum, the early concessions provided for a royalty to be paid in money but calculated on the basis of physical output. Thus, for example, the Turkish Petroleum Company Concession (1925) in Iraq provided for payment by the company to the government of a fixed royalty of four shillings (gold).

The basic advantage of a royalty is that it is a guaranteed payment to the government for the depletion of the resources irrespective of whether the company makes a profit or not. Further, a royalty payable on the basis of physical output (that is, volume of production or shipment) is easier to administer than a royalty based on value, or a tax on profits, since this gives rise to complex problems, such as the adequacy for purposes of valuation of crude oil of the realised price, which in most cases is a transfer price paid by an affiliate, or the rationale and admissibility of deductions from profits for purposes of assessing tax.

A royalty, however, by its nature, being a payment made irrespective of profits, is not an effective instrument for capturing the entire rent or surplus generated by petroleum operations. Since royalties are paid out at an earlier point in time than profits, they are a marginal cost to the company. In other words, the higher the royalty the greater is the cost to the company, which means that royalties beyond a certain level can render operations unprofitable, leading to premature abandonment of partly developed fields or to negative decisions with regard to the development of marginal fields.

A variety of devices has been developed to reduce the deterrent effect of royalties, and to so design them as to provide certain types of incentives to companies. Thus, in Britain, the government has been empowered to remit royalty in cases where it is judged that payment of royalty would render a project unprofitable. It has, however, been observed that:

> The difficulty with royalty remission as a technique is that it must inevitably be the result of a negotiation in which the parties could be expected to engage in a good deal of strategic behaviour and in which therefore the outcome would not necessarily be uniformly optimal. The greater the royalty, the more often remission would have to come into play. At the limit, if an attempt were made to capture all of the economic rent through royalties, the royalty schedule would have to be separately negotiated in each case.[51]

A different technique to introduce flexibility in a royalty system is that of sliding-scale royalties, such as those adopted in Canada and in Norway. The operation of sliding-scale revenues in the western provinces of Canada is described thus:

> . . . a sliding scale applied to each producing well rather than a fixed royalty on oil is prescribed. An obvious purpose of the sliding scale is to maximise resources in the case where a well has a high rate of production, and so royalties as high as sixteen and two-thirds per cent may be payable. A less obvious purpose, but one that has had more application in areas where well spacing is relatively close owing mainly to fragmented ownership, is to prevent wells from being prematurely abandoned owing to low producibility. To accomplish this purpose royalties may be as low as five per cent.[52]

The Norwegian sliding-scale royalty prescribed by 1972 Royal Decree was as follows:

Average field production allowed	Royalty
40,000 barrels per day	10%
100,000 barrels per day	12%
225,000 barrels per day	14%
350,000 barrels per day	16%[53]

The sliding scale is thus constructed so as to require a lower rate of royalty from small fields, which are viewed as being marginal. A critique of the sliding-scale royalty mechanism, however, argues that:

The underlying assumption is, of course, that large fields are more profitable than small fields, but this assumption does not always prove to be true in practice for two reasons. First, large fields can be marginal, particularly where they lie in deep water or far from shore or where there are great technical risks associated with the particular geologic structure. Second, large fields must be developed in conjunction with large pipelines, which have to be constructed early in the project's life, whereas small fields could possibly be handled by tanker loading with the consequence that transportation costs would be spread more evenly over the lifetime of a project. In that event, a small field may be more profitable . . . [Further], sliding scale royalties may be complex to administer in practice, since through their progressivity, they create a disincentive to economically optimum rates of production [the more so], as it should be borne in mind that the optimum rate of production from a field is not only a matter of judgment, but also changes throughout the lifetime of a project — building up rapidly as development wells come into production and then trailing off gradually . . . with perhaps some temporary increases as secondary and tertiary recovery investments are made.[54]

Royalty in recent agreements is expressed to be payable "in cash or in kind" . . . in terms of a percentage of "the total quantity of crude oil produced" (Egypt-Esso Agreement of 1974), or "as a percentage of the Posted Price of Crude Oil" (Abu Dhabi-Maruzen Oil Co. (and others) Agreement of 1967). Such royalty payments were to be made quarterly.

The share of crude oil made over to the government or the national oil company, under a production-sharing contract, is in effect a royalty payment in kind. It is noteworthy that a number of recent production-sharing contracts, considered in the preceding chapter, have introduced sliding scales for determining the share of production to be made over to the national company, as for example the Phillips-Pertamina PSC of 1974 and the Bangladesh PSCs of 1974.

It is important to note that while under country tax regimes, such as that of the United States, a tax credit is available in respect of income tax paid by the company overseas, such a credit is not available in respect of royalty payments. The position of the US Internal Revenue Service, discussed in the preceding chapter, was that the share of production paid to Pertamina was in substance a royalty and that even though a part of it may have been paid by Pertamina to the Indonesian treasury to meet the income-tax obligations of the oil company, no part of such royalty was identifiable as an income tax or a tax in lieu of income tax and therefore the oil company was not entitled to a foreign tax credit. The IRS ruling, discussed in the previous chapter, enunciates criteria by reference to which a payment to a foreign government may be characterised as an income tax or royalty. It attaches more importance to substance, than to form. Indeed on the basis of these criteria, the Canadian Progressive Incremental Royalty may well be treated as an income tax in respect of which a foreign tax credit might be admissible in the United States.

Taxation

Royalties, while providing a steady and assured payment based on actual production to the state as the owner of the resource, are not effective as a mechanism for securing for the state a substantial part of the profits generated by the extraction and sale of the resource. The reason is in part inherent in the character of a royalty, for as has been observed:

> Clearly, firms have been reluctant to take on heavy royalties. From the company's point of view, a commitment to a large royalty particularly in the early years of an extractive operation is particularly dangerous. At the outset, the firm faces a great deal of uncertainty as to whether it will be able to extract the natural resource profitably. The case of the royalty represents to the firm an additional cost of extraction, one that will be incurred whether the project is profitable or not. On the other hand, a commitment to pay an income tax on profits if they do materialise appears less risky. If there are no profits, the company has no obligation to pay tax to the host government. Under a pure income tax arrangement, the firm incurs significant obligations only if profits are high. With a drive to avoid risk, the foreign firms have usually been willing to agree to an income tax that, if the expected level of profits results, would be larger than any payments that would be agreeable under a royalty arrangement.[55]

Governments have, therefore, increasingly resorted to income taxation to appropriate a share of the profits arising out of petroleum operations. The fifties witnessed the acceptance of the "fifty-fifty" profit-sharing formula, which was implemented in most cases through the introduction of a system of income taxation.

The "fifty-fifty" formula has progressively been altered in favour of the state. A rough calculation, made by the British Department of Trade and Industry, of the percentage of profits appropriated through a royalty-cum-tax system in different countries in 1973, indicated as follows:

United Kingdom
    12½ per cent royalty on well-head value;
        Corporation tax at 50 per cent   55–60

Middle East
    12½ per cent royalty, tax at 55 per cent, both
        based on posted price   75–80

Nigeria
    10 per cent royalty, tax at 55 per cent, both
        based on posted price at export terminal   70–85

Norway
    (a) Terms up to December 1972
        10 to 12½ per cent royalty on agreed price,
        tax at 50.6 per cent   55–60
    (b) Terms just announced for future licences
        Sliding scale royalty of 8 to 16 per cent on a
        price to be determined, tax at 50.6 per cent,
        carried interest up to 50 per cent   55–80

Netherlands
    0 to 16 per cent royalty on well-head value,
        tax at 48 per cent   50–60

USA
    16⅔ per cent royalty, corporate tax at 48 per cent
        of profit less depletion allowance   50–60[56]

A high rate of tax, however, does not by itself necessarily ensure a high government take, in particular in the early years of a project, as a result of the capital allowances and other deductions which companies are entitled to make from their gross income. Provisions regarding capital allowances and deductions therefore merit careful attention as they are as important as the tax rate in determining the magnitude of the government take and the profitability of a venture from the company's point of view.

Government take can be substantially reduced by certain types of allowances and deductions, which have subsequently been perceived by governments, or their legislatures, as "loopholes". Thus, the British Public Accounts Committee, in its Report in 1973, identified a number of such gaps in the taxation rules. These included provisions which allowed deductions in respect of (a) substantial capital allowances; (b) losses made from transactions in oil at artificial posted prices (a company could show a loss by pricing the oil purchased from a producing affiliate at a transfer price, which was higher than the market price, the effect of which was to increase the production companies' profits while the profits of the marketing company were correspondingly reduced); (c) group losses, that is losses made by a group or consortium of companies in transactions unconnected with operations in the British sector of the North Sea, could be set off against group profits in Britain. Thus the Report observed that:

> If the present situation continued . . . the U.K., could expect little tax revenue from profits arising on continental shelf oil production even by 1980; annual capital allowances would neutralise a large part of revenues. Middle East tax losses accrued and accruing would in many cases take the rest unless these losses had been absorbed in the meantime.[57]

The 1974 White Paper on UK Offshore Oil and Gas Policy proposed a number of measures to close these loopholes. These included the cancellation of losses accumulated as a result of overseas transactions, strengthening the existing transfer pricing legislation in the Income and Capital Taxes Act, and defining a "ring fence" around the North Sea for tax purposes. The rationale for the "ring fence" is explained in the White Paper in the following terms:

> There is the question of what has come to be called a ring fence for the North Sea . . . Exchequer receipts from the North Sea should not be at the mercy of allowances and losses resulting from extraneous activity . . . group relief, therefore, should not be allowable against profits from North Sea activities. Correspondingly, North Sea profits of a single company with other activities will not be reduced for tax purposes by losses or allowances arising from those other activities. Nor will a company with losses or excess allowances be able to use them to claim payment of the importation tax credit on dividends from a company within the group and paid out of North Sea profits.[58]

Government take in the early years could be substantially reduced due to the operation of capital allowance provisions and the time lag between the receipt of monies and the payment of tax. Thus, it has been estimated that

> A North Sea company could recover nearly all development costs before any corporation tax would be due. And if the company then were to develop another field the capital allowances applicable to that development could further defer the onset of corporation tax payments.[59]

The time lag of taxes itself provides a higher rate of internal rate of return, as the example below illustrates:

> A simplified example may help to illustrate the effect of taxation on the internal rate of return. Assume that all revenues from a project flow in at the end of year three and that they amount to $1,000 million. Ignoring the capital allowances point, assume that an 80 percent tax is imposed, but that it is not payable until the end of year five. At a 40 percent rate of discount, the $1,000 million receipts have a net present value at the time of initial investment of $364 million whereas the $800 million of tax has a net present value of only $148 million. Thus, at present value at the time of the decision to go forward, the 80 percent tax would amount to much less than half of the revenues.[60]

The administration of a system of income taxation presents a whole range of complex problems. Among these are: determining the sources of income that are to be taxed, the prices that would govern the sales made by the companies, the application of provisions relating to capital allowances, and the calculation of costs and expenses and other deductions that are chargeable against gross income.

Gross income. Since multinational oil companies operate globally, rules need to be adopted to determine the part of the income of the company which will be taxable in each of the countries where it carries out operations. It is inherent in the character of multinational firms that they can organise their affairs in such a way that income can be shifted to affiliates, out of the reach of the tax regime of the country in which operations are being carried out, if they aim to avoid host country tax obligations. To safeguard this, governments must be prepared to monitor complex transactions, and to adopt rules which can be effectively administered to check schemes designed to avoid host country tax obligations.

Sale price. The problem which presents considerable difficulty to the tax administration is that of determining the price to govern the sales of the petroleum. In the petroleum industry, a large part of the sales of oil companies are to their affiliates, at "transfer prices", which are lower than those which could be obtained in arm's length transactions. Governments have resorted to different mechanisms for dealing with the problem. In the traditional producing countries in the Middle East a system of "posted prices" was established, whereby taxes were computed on the basis of a price, which was published, or "posted" periodically by the companies. The subsequent developments which ultimately led to governments determining what "the posted price" should be has been described in Chapter I. In Venezuela, where taxes were assessed on the basis of realised prices, a Commission inquiring into sale prices in 1960 found a number of cases where Venezuelan crude oil had been "sold" at prices lower than those prevailing in the market, and that abnormal discounts were being allowed. This led to pressure for a change in the rules. Thus in 1966, a new system was introduced whereby all crude oil and petroleum products exported from Venezuela for a period of 5 years (1967-71) would be assigned "reference prices" for the purpose of taxation. These would be negotiated as agreed in advance for the whole period. Should realised prices be above the reference levels, taxes would be paid according to the higher prices.[61] In Norway, the Finance Committee of the Storting, recognised that "most of the crude oil which is sold is sold between units within the integrated international oil companies", and thus that "a substantial share of the trade in crude oil is thus not subject to ordinary price mechanisms". It therefore proposed that, for purposes of taxation, petroleum should be valued at a "norm price" to be determined by the Petroleum Price Board, which would periodically determine the "norm price" on the basis of material and information collected by it and of special studies, and after considering statements and materials, as well as views, submitted by the companies.

PSCs provide for valuation of the oil on the basis of realised prices. They do, however, incorporate mechanisms to deal with sales to affiliates in the form of provisions which: (a) prohibit the payment of commissions or brokerage to affiliates; (b) require sales to affiliates to be valued "at the negotiated average net realised price f.o.b. point of export determined in respect of other sales (excluding any sale which is on terms not competitive with or comparable to arm's length purchases and sales between commercial buyers and others); and further provide that (c) where no such other sales have occurred, the price is to be determined "in a commercial manner taking into account prices at which comparable types and quantities have been sold in competing export markets, bearing in mind in that connection possible differences in quality and in transport costs".

Capital allowances. The principal capital allowances which have been considered above are: depreciation, amortisation and depletion. Depreciation representing the deterioration in value of assets is charged against profits. As indicated above, the amount that can be charged is affected by what is taken to be the assumed life of an asset, the rate of depreciation allowed, and the method of depreciation applied, such as the straight-line method or the declining-balance method. Thus, government take in the early years can be substantially reduced and the DCF rate of return for the company substantially increased by allowing the company to assume a shorter presumed economic life of assets so as to allow it to be written off more quickly, by allowing a higher rate of depreciation, and by employing the declining-balance method. Amortisation, in the sense of the write-off of capitalised expenditures, that do not represent fixed assets (and not in the sense of repayment of a loan or debt) is another item chargeable against gross income. Problems have arisen with regard to items of expense which may be amortised in particular in relation to expenses incurred before the agreement and expenses incurred outside the country. Thus recent agreements have specifically spelt out items of expense which cannot be treated as "deductible expenses"; for example, the Abu Dhabi-Maruzen Agreement of 1977 specifically provides that the following among other items could not be deducted for purposes of determining taxable income: (a) foreign taxation paid on income derived from sources in Abu Dhabi; (b) interest or other consideration paid or suffered by the company in respect of financing its operations in Abu Dhabi; (c) expenditure in relation to the organising and initiating of petroleum operations in Abu Dhabi; (d) bonuses and rentals paid to the Ruler.

Depletion is an allowance premised on the assumption that the oil discovered constitutes a capital asset, and since a portion of the oil extracted is irreplaceable, an allowance should be permitted as compensation for the loss of the capital value of the irreplaceable oil. Originally, depletion allowance had been limited to the cost of the total investment so that once the cumulative allowance equalled the investment no further deduction was permitted. In the United States, under the pressure of the industry, a new basis was enacted, so that a deduction of 27½ per cent could be made from the gross income by way of depletion allowance. The effect of percentage depletion calculated on the basis of gross profits is to substantially reduce the tax payable to the state. Thus, a calculation made on the basis of data relating to twenty-seven oil companies (for the period 1945-54) indicated that as a result of the allowance of percentage depletion, oil companies were paying corporate income tax at the rate of 17 per cent while other enterprises were subject to a 52 per cent rate.[62] While a percentage depletion allowance may operate as a tax incentive there is a considerable preponderance of views that this type of incentive is likely to prove too generous, since it is difficult at the outset to assess the levels of profit that are likely to be attained throughout the life of the project.[63]

Operating costs. In assessing tax, operating costs are deductible from gross income. Problems are posed in this area in particular by transfer pricing, that is prices charged by affiliates, for goods or services supplied by them. Just as in the case of the sale of oil to affiliates the problem arises of artificially deflating the sales price, the problem arises when goods or services are purchased from affiliates, as is often done, of the artificial inflation of prices. The monitoring of prices of goods and services purchased for the purposes of petroleum operations is a complex problem; in particular where inter-affiliate transactions among subsidiaries of multi-national corporations are involved.

There are different mechanisms which have been devised to provide some protection against over-charging. Thus a rule may be prescribed that charges for goods and services supplied by affiliates should not be greater than the charges that would have been incurred if these were acquired from a non-affiliate. Alternatively, a cost-plus formula may be adopted, where the prices that should be allowed for taxation purposes is the cost, plus a specified margin. Some of the recent agreements, such as the Egypt-Amoco Agreement of 1976, have incorporated a mechanism whereby costs are monitored on the basis of quarterly reports; in the event of an objection raised on an item, which cannot be amicably resolved, there is provision for the matter to be referred for expert determination.

The problem of monitoring of costs is just as taxing under PSCs, since costs have to be determined for the purpose of appropriation by the company of a proportion of the oil as "cost oil", to be applied towards cost recovery. The higher the total costs, the greater the amount of oil that the company would be entitled to appropriate under the head. It has therefore been observed that

> The costs of operations, although limited to 40 per cent, must be calculated to determine the amount of oil that goes to each party . . . Slippage in the amount of income accruing to the government could occur in the calculation of these "operating costs" incurred by the company under post-1965 agreements. Such deductions must be given the quality of scrutiny that would be given by a government tax office to deductions from gross income in a traditional concession agreement.[64]

The PSCs, however, are so structured that in the early years they provide a higher government take than under the other types of arrangements which provide mechanisms for capital allowances and cost recovery, such that the government take in the initial stages of the project is substantially reduced. The PSCs, by limiting cost recovery to a stipulated percentage of the oil (40 per cent in the standard PSC, 20 to 30 per cent in the more recent PSCs) produced each year, leave the balance of the production to be shared in agreed ratios. An analysis of the financial mechanism embodied in the PSC demonstrates how it operates to the advantage of the state in the early years, thus:

> Royalty provisions were designed to ensure that the host country obtained a fixed financial return irrespective of the profitability

212

of the venture. The PSC's have effectively achieved this result because Indonesia receives a higher guaranteed income during the initial stages of production, when this venture is not operating at a profit. Under the typical concession agreement, in the early stage of the operation all costs came off the "top" of production; the host country receives a modest royalty payment but no income tax revenues. In Indonesia, costs are amortised such that the minimum payment to the government is never less than 39 per cent.[65]

Participation

Governments have increasingly made provision for state participation in petroleum development arrangements. The different forms of such participation have been discussed in the preceding chapter. Participation clearly has financial implications — these implications, however, vary from one type of participation to another. As a participant in the equity, the state in addition to royalty and taxes becomes entitled also to a dividend. It has, therefore, been observed by one analyst that participation "is an ingenious way of further increasing the tax per barrel without touching either posted prices or nominal tax rates".[66] Some critics of participation have maintained that it is simpler to raise the government take by other mechanisms than equity participation, since the latter involves capital outlays by the government itself. In cases where governments have forgone their right to levy an income tax in exchange for a 50 per cent equity holding without any financial contribution, it is pointed out that

> In general, holding 50 per cent of the equity is, in purely financial terms, less attractive to the government than is an income tax at a 50 per cent rate. Under the ownership arrangement, the government receives half the dividend payments. But half of the dividend payments is usually less than half of the taxable profits of the enterprise.[67]

The UK Fifth Round licensing terms (early 1977) were remarkable in that they carried the logic of participation a stage further so that the burden of costs was shared by the government at all stages, including exploration. Thus, it was provided that not only would the British National Oil Corporation (BNOC) be a 51 per cent partner in any licence (except where the British Gas Corporation was a licensee) and thus be entitled to a 51 per cent share in all the benefits of the licence, including a proportionate share of the petroleum produced and the income accruing from discoveries within the licensed area, but that BNOC would also be responsible for its share of all expenditure and that it would pay its share of expenditure for exploration, appraisal and development work as it was

213

incurred. Provision was, however, made for BNOC to elect not to participate in a development or exploration venture. In that case, the company
could undertake such a venture at its *sole risk* and *cost*. While BNOC
would be entitled to vote in the operating committee, appropriate to its
having a 51 per cent share and was eligible to be appointed the operator,
it could not exercise its majority vote to appoint itself the operator.

The Sixth Round licensing terms (1978) represented a further
improvement of the terms from the point of view of the government. The
participation provisions were also more favourable to the government.
Thus, operators were invited to make offers in respect of meeting some or
all of BNOC's share of exploration and appraisal costs; agreeing to BNOC's
share of equity being higher than 51 per cent and to BNOC having the
option to purchase/sell some or all of the oil produced. It was provided
that these offers were to be taken into account in the assessment of the
applications. A further criterion introduced for judging applications was:
the continued satisfactory performance of the applicant in accepting and
implementing majority state participation in existing licences.

As was evident from a consideration of different types of participation
in nearly all the cases the risk and investment in exploration is borne by
the company, and it is only upon discovery that the government has the
option to "buy-in" to the project upon assuming a proportionate share of
exploration and development costs (though in some recent agreements, as
in Norway, the state is not required to reimburse the company for a proportionate share of exploration costs). In theory, where the government
does not reimburse the company for a part of the exploration costs, this
could have a deterrent effect on exploration. In practice this is not the
case since exploration costs are very small compared to the scale of
development costs. Therefore, even if the government does not assume a

part of the exploration costs, but shares in the development costs, there
may be no adverse effect on the internal rate of return for the company.
Indeed there may be cases where the government may be able to obtain
development capital at a lower cost than the companies. This would be
particularly true in the case of developing countries, which may have
access to development capital from international institutions, such as the
World Bank, at a lower cost than that at which it would be available to
companies.

In countries such as the Netherlands, certain financial advantages are
conferred on companies in the case of projects where government participation is involved. Thus in cases where the Dutch Government undertakes

to participate, the royalty charged was only 50 per cent of the regular royalty (though the extent of government participation was 40 per cent) and the sliding-scale royalty stopped at 8 per cent instead of going up to 16 per cent as was applicable for large fields.

It is, however, observed that:

> To emphasize financial techniques for offsetting any financial disincentive effect of government participation, however, obscures the central point that the primary disincentive comes from the control element of government participation.[68]

While companies may not favour government participation, the evidence world wide is that they are reconciled to it, and have learned to live "and operate" with it. The arguments adduced in support of participation in Britain by one of its leading protagonists are:

> Participation is as good an investment for government as it is for companies . . . The second argument in favour of participation is that it automatically provides a check on investment outlays and production costs of the oil companies – it [also] helps to provide security of supplies.[69]

Bonuses

(a) *Signature bonus*. Many of the recent agreements provide for payment of a fixed sum, by way of bonus, upon signature. The amounts paid by way of signature bonus have not been insubstantial, ranging upward from one million dollars. A criticism of this form of payment is that

> a fixed signature bonus can have a disincentive effect, since unlike an auction bid, it is unrelated to the economic rent expected to be derived . . . it [therefore] discourages companies from seeking licences in marginal areas and is ill-designed to capture economic rent.[70]

The signature bonus is clearly not an instrument for capturing rent since it is a one-time payment made right at the outset, even before the commencement of exploration. A signature bonus can, however, serve much the same purpose as an auction, if it is not fixed, and provision is made for bonus bidding, so that competing companies are invited to bid amounts which they are willing to pay as signature bonus, above a minimum stipulated amount. The advantages of this type of "bonus bidding" have been discussed in Chapter III above. It is also not correct as has been suggested that a fixed signature bonus is "functionally equivalent" to an area rental,[71] since as will be suggested below, area rentals principally serve the function of discouraging the practice of "sitting on" concessions.

(b) *Commercial discovery bonus/production bonus.* Many of the recent agreements provide for a one-time payment of not insubstantial amounts, upon the making of a commercial discovery, and upon certain stipulated levels of production being attained. Such lump-sum payments constitute welcome contributions to the foreign exchange receipts of countries, specially those which suffer from foreign exchange shortages. Being one-time lump-sum payments, however, they are not economically significant, and given their magnitude, they are unlikely materially to affect the profitability of a venture from the company's point of view.

Area/surface rentals (and mandatory relinquishment/surrender provisions)

Under most arrangements some provision is made for area or surface rentals to be paid to the state. While under the earlier arrangement the rental payments tended to be fixed, and did not involve large amounts, under a number of recent arrangements, sliding-scale surface rentals have been introduced, to operate as an instrument to discourage the retention of areas over which exploration is not being conducted, or conversely, to exert financial pressure to relinquish such areas, so that these may be allocated to others who may be keen to conduct explorations in these areas. Thus, the Norwegian legislation provides for a steeply escalating surface rental, which was recently further increased with the specific object of making it prohibitively expensive to retain areas in which companies were not willing or able to undertake substantial programmes of exploration. A similar sliding-scale device was incorporated (as was pointed out in the preceding chapter) in the Indian PSCs concluded in 1974. In these cases, sliding-scale surface rentals could be said to be functionally equivalent to mandatory surrender or relinquishment provisions, such as are contained in most recent petroleum development arrangements.

It should, however, be noted that mandatory surrender provisions, in addition to being used as instruments for accelerating exploration and providing for early relinquishment of areas over which exploration was not being carried out, has been used for the distinct purpose of enabling the state to "share the fruits of discovery", as in Alberta and Saskatchewan. This use of such area reduction (or mandatory surrender) provisions is said to "mark off the Canadian legislation from that in other parts of the world", and its effect is described thus:

> Area reduction provisions may have two objectives — one is to avoid long-term monopoly holding by encouraging the entry of new

operators for a second round of exploratory activities; the second is to give to the state a portion of the fruits of discovery. In most parts of the world, and in the Australian states, the relinquishment provisions are directed primarily to the first objective. In the western Canadian provinces they are directed to both. The proposed offshore code in Australia has borrowed from the Canadian precedent, for its two-stage relinquishment scheme has as the object of the first stage, the opening up of a permit area to new operators, and, as the object of the second stage, the sharing by the Crown in the benefits of the permittee's discovery.

The Canadian area-reduction scheme, based on the Alberta example, accomplishes this second objective by requiring the permittee in the seventh or eighth year of his permit or reservation, if not sooner, to "go to lease" on not more than fifty per cent of the acreage under a selection system that requires him to choose leases in square shapes of not more than nine square miles or in rectangular shapes of not more than eight square miles either on a checkerboard pattern or on a pattern with one-mile corridors between the leases. The remaining acreage then comprises "Crown reserves" to be disposed of under a bid system either as Crown Reserves Drilling Reservations or as leases.[72]

The Australian Common Code, embodied in the Petroleum (Submerged Lands) Act, 1967, provided for a similar "checker-board" system of relinquishment whereby the company, upon discovery of oil or gas in commercial quantities, was required to yield four-ninths of the location to the state, which would be free to sell the area relinquished to the highest bidder. This was one of the features which had led the industry to characterise the Common Code as "the most restrictive offshore legislation in the western world".[73] Their concerted representations to the government proved to be effective in that an alternative provision was made whereby the company could exercise a pre-emptive right to acquire the four-ninths by paying an additional royalty over the entire area of between 1 to 2½ per cent.[74]

Provisions to benefit the national economy

Most recent agreements contain certain provisions, having financial implications, intended to benefit the national economy. Among those which are most common are provisions which require: (a) training of nationals; (b) purchase of goods and services locally; (c) supply of oil or gas to meet local requirements; (d) establishment of refinery or other processing plants, or other industries.[75]

Training obligations are generally assumed under most types of arrangements, and under a number of them, in particular the PSCs, companies undertake to contribute towards the cost of training and the setting up of training establishments. The benefit conferred on the host country in such cases is to be seen not just in terms of the financial gain resulting from the cost of training being borne by the company, but more significantly in terms of the transfer of technology and expertise that can be effected through properly formulated training obligations being imposed on companies.

The obligation to purchase goods and services locally is usually qualified by the provision that preference should be accorded to their procurement only if they are of the requisite quality and are available at prices which are competitive. If this proviso were not there, the subsidy to local suppliers would be borne as an additional cost by the company, which could in certain cases affect the profitability of the venture.

The obligation to meet the domestic requirement, or a part thereof, is contained in many of the recent agreements.[76] As indicated above, this can substantially affect the profitability of the venture, unless the supply is to be made against payment in foreign exchange, at international market prices.

Finally, there are provisions which seek to impose an obligation on the company to invest in the establishment of refineries or industries connected with petroleum. In most cases, such obligations are made subject to the economic feasibility of such investment being established and upon terms which are to be negotiated and agreed. Clearly the imposition of unqualified obligations in this respect, which might subject the company to an open-ended liability to make investments of an indeterminate magnitude, could render a package, of which it was a part, unacceptable to a company.

APPENDIX A[77]

While making allowances for the difficulty in providing a comparative assessment of the financial terms on the basis of the calculation of the DCF rate of return, an exercise carried out on the basis of the assumptions spelt out below gives some indication of the relative attractiveness in financial terms of different national arrangements (Britain, Norway, Canada, Egypt, and Indonesia):

Assumptions:
Price: $12.50 per barrel.
Size: 300,000 barrels per day, which reaches its maximum level for about 3 years and produces 1 million barrels during its 20-year life.
Development costs: Table A is based on costs of $4000 per daily barrel, and Table B on costs of $8000 per daily barrel.
Operating costs: Total operating costs over the life of the field has been assumed to be equal to 80 per cent of the development cost.
Financing: Prices and costs have all been computed in 1976 terms. Companies have been assumed to finance the operations from their own resources and where state participation is provided for on a carried interest basis, the rate of interest payable to the company on the outstanding cash balance has been assumed to be 10 per cent (the interest to be taxable at source).

Summary of terms: (country-wise)
BRITAIN
Royalty: 12½ per cent of the tax value (i.e. $12.50 per barrel for the purposes of the present analysis).
Corporation tax: 52 per cent (tax payment is deferred on year).
Petroleum Revenue tax (PRT): 45 per cent. Income subject to PRT is gross revenue (less the value of 500,000 tons of petroleum for each 6-monthly period, up to a maximum of 1 million tons during the life of a field) less operating costs, royalty and capital expenditures (plus an additional uplift of 75 per cent); PRT is limited by a safeguard provision such that if in any year the income subject to PRT is equal to or less than 30 per cent of total capital expenditure to date PRT tax is cancelled; PRT is also limited by a tapering provision such that the PRT charge is not greater than 80 per cent of the amount (if any) by which income subject to PRT for the year exceeds 30 per cent of capital expenditures to date. PRT is allowable as a charge against corporation tax.
Depreciation rules: 100 per cent annual write-off allowed for the purpose of corporation tax computation, but an additional uplift of 75 per cent

219

of expenditures is allowed for the purpose of PRT, also at 100 per cent write-off.

Participation: It has been assumed that the government (BNOC) takes a 51 per cent share in the hypothetical field. In one case, it is assumed that the state pays its share of development costs as they are incurred; in the other case it is assumed that the state participates on a carried interest basis.

NORWAY

Royalty: up to 39,999 b/d 8 per cent
40,000–99,999 b/d 10 per cent
100,000 –224,999 b/d 12 per cent
225,000–349,000 b/d 14 per cent
350,000 b/d and over 16 per cent

If production from the field is produced over a 50-day period the royalty is reduced in accordance with the table, but not to a lower rate than 12 per cent. The value on which royalty is determined is negotiated between the government and the companies. If agreement is not reached, the government determines the value, taking into account market prices for North Sea crude. It has been assumed that the royalty valuation equates to the market price of $12.50 per barrel.

Corporation tax: 50.8 per cent (tax payment is deferred one year).

Special tax: 25 per cent, which is additional to corporation tax. This tax will be calculated on gross revenue, less operating costs, depreciation, royalty, plus 10 per cent of the acquisition cost of all equipment put into operation in the previous 15 years.

Depreciation rules: $16\frac{2}{3}$ per cent annual depreciation on a straight-line basis. Applicable on both corporation and special taxes.

Participation: No maximum set down, but it has been assumed 50 per cent in the present calculations. In one case it has been assumed that the state pays its share of development costs as they are incurred: in the other case it has been assumed that the state participates on a carried interest basis.

CANADA

Royalty: 10 per cent (it has been assumed that royalty is valued on the basis of the well-head price).

Progressive incremental royalty (PIR): Fields will be subject to an incremental royalty, which will not apply in any year unless annual profits, divided by depreciated ring-fenced investment, are greater than 25 per cent. As profitability increases, the rate of tax also increases, although at a slower rate. The maximum rate of tax is 40 per cent. For the purpose of the PIR formula ring-fenced investment is depreciated at a rate of

10 per cent per annum on a declining balance basis. The annual profits for PIR are calculated as follows: revenues less operating costs, capital allowance, federal royalty and taxes. Depletion allowance is not allowed as a deduction for PIR purposes.

Federal and provincial income tax: 46 per cent.

Capital allowance: This provides for the write-off of investment at the maximum rate of one-sixth per year starting in the year of investment. If allowances are not required to reduce or eliminate the PIR payable they may be carried forward and used at the discretion of the operator in later years.

Depletion allowance: An allowance of $33\frac{1}{3}$ per cent of net income before tax may be charged against taxable profits provided that $3 is spent on operations for each $1 depletion allowance (it has been assumed that the company claims the allowance for around $\frac{1}{3}$ of its capital costs on developing the field, but that it claims no further allowance because it does not continue exploring in Canadian territory).

EGYPT

Production share: In recent agreements it has been split 80/20 in favour of the state.

Costs: Costs may be recovered by the contractor (company) out of up to 40 per cent of production, any balance unrecovered to be carried forward for recovery in subsequent years. After deduction of production to cover costs the proceeds of the balance of cost recovery crude are to be handed over to the state company.

Contractor's taxation liability: Payable by state company out of its share of production at no cost to contractor.

Depreciation: 20 per cent per annum, for exploration expenditure, and 10 per cent per annum for development expenditure.

INDONESIA

(a) *Pre-1976 terms Production share.* Most contracts were split 65/35 in favour of the state; with provision for graduated increases up to 70/30. The latter figure was reached when production increased to 75,000–200,000 barrels per day, according to contract. Each company had to provide crude oil for domestic consumption at a price of cost plus 20 cents per barrel, in proportion to its share of total Indonesian production.

Costs: Recovered by contractor out of up to 40 per cent of production; any balance unrecovered to be carried forward for recovery in subsequent years. After deduction of production to cover costs, the balance of production was split as above.

Surtax: Surtax at the rate of 85 per cent was payable on the income arising as a result of the difference between the sales price and $5 per barrel. The $5 per barrel base price was subject to escalation to reflect inflation.

Contractor's taxation liability: Payable by state company out of its share of production at no cost to contractor.

Rental: Contractor (company) pays a rental per annum, not exceeding 10 per cent of cost, on movable physical assets purchased by contractor (company).

Depreciation: To the extent allowed by maximum 40 per cent production set aside to cover costs.

Although broadly similar in content, Indonesian participation sharing terms differed from each other in certain respects, especially as regards the graduation of the division of the production share according to the level of production.

(b) *The latest terms*: The new terms are as follows: an 85/15 split (after allowing for the recovery of costs) in favour of the state. Out of the contractor's share, he must sell 20 per cent to the government at cost plus 20 cents per barrel for local sales (i.e. the contractor only obtains the full market price of $12.50 per barrel on 80 per cent of his 15 per cent share).

Cost recovery: Contractors with existing production are divided into two groups. Group I are those with proved recoverable crude reserves of less than 7 years. Group II are those with proved recoverable reserves of 7 or more years. Costs are recovered as follows:

Group I: Costs unrecovered on 1 January 1976 (costs incurred before, or prior to starting production if later than, 1 January 1976): capital costs will be depreciated over an average of 7 years using the double declining-balance amortisation method. Non-capital costs will be depreciated over an average of 5 years using the straight-line method, but the contractors will be allowed 8 per cent interest on the annual unrecoverable balance.

Costs incurred after 1 January 1976: capital costs will be depreciated over an average of 7 years on the basis of half the depreciable life, using the double declining-balance method. Non-capital costs will be expensed.

Group II: Costs unrecovered on 1 January 1976 (definition same as for Group I): Capital costs will be depreciated over an average of 14 years using the double declining-balance method. Non-capital costs will be depreciated over an average of 10 years using the straight-line method, with 8 per cent interest allowed on the annual recoverable balance.

For yet-to-be-developed fields, the depreciation rules are an immediate write-off for 30 per cent of costs, with the remainder depreciated at $6\frac{3}{4}$ per cent per annum.

Table A

FINANCIAL EFFECTS OF SELECTED CONCESSION TERMS

($4000 per daily barrel)

Concession terms	Undiscounted Averages $ per barrel of crude produced			Government's share of profit %			Company's approx. DCF rate of return
	Costs	Company margin	Government revenue	Undiscounted	Discounted @ 10%	Discounted @ 15%	
United Kingdom*	2.048	1.518	8.934	85.5	85.9	86.7	36.3
United Kingdom†	2.048	1.577	8.875	84.9	87.0	89.4	25.7
Norway*	2.048	1.265	9.187	87.9	87.6	88.3	33.6
Norway†	2.048	1.325	9.127	87.3	88.6	91.0	24.2
Egypt	2.048	1.500	8.952	85.6	93.6	99.8	15.2
Indonesia‡	2.048	2.292	8.160	78.1	82.1	85.9	25.2
Indonesia §	2.048	1.642	8.810	84.3	91.8	97.3	17.1

* State participation assuming no carried interest.

† State participation assuming carried interest.

‡ Pre-1976 terms.

§ New terms.

Table B

FINANCIAL EFFECTS OF SELECTED CONCESSION TERMS

($8000 per daily barrel)

| Concession terms | Undiscounted Averages $ per barrel of crude produced | | | Government's share of profit % | | | Company's approx. DCF rate of return |
	Costs	Company margin	Government revenue	Undiscounted	Discounted @ 10%	Discounted @ 15%	
United Kingdom*	4.095	1.512	6.893	82.0	85.1	89.6	20.4
United Kingdom†	4.095	1.685	6.720	80.0	89.2	100.0	14.6
Norway*	4.095	1.213	7.192	85.6	89.6	95.3	17.5
Norway†	4.095	1.391	7.014	83.4	93.8	100.0	12.7
Canada	4.095	2.727	5.678	67.6	74.0	83.9	19.2
Egypt	4.095	1.370	7.034	83.7	100.0	100.0	8.5
Indonesia‡	4.095	2.266	6.138	73.0	88.1	100.0	13.8
Indonesia§	4.095	1.745	6.660	79.2	100.0	100.0	10.0

* State participation assuming no carried interest.
† State participation assuming carried interest.
‡ Pre-1976 terms.
§ New terms.

EXTRACTS FROM GOVERNMENT OF PAPUA-NEW GUINEA'S
STATEMENT ON PETROLEUM POLICY AND LEGISLATION
(MARCH 1976)

Taxation and financial regime

A. *Summary*

In the determination of an appropriate financial regime the government has been constantly aware of the need to look at all fiscal aspects together as one package. It is not appropriate to examine the severity of the proposed taxes or the participation provisions in isolation since they all impact upon the economics of petroleum operations. In selecting appropriate terms a major consideration has been to provide terms attractive enough to ensure a continuing high level of exploration in the immediate future, yet flexible enough to assure the state a high proportion of profits from very profitable fields. Considerable research has been undertaken into recent fiscal provisions of other countries containing areas of unproven oil potential and high costs of development. In many areas the oil companies have accepted terms which for an "average" size field provide a (discounted) profit split of 70-75 per cent to the state and 25-30 per cent to the company. It also appears that the minimum real rate of return acceptable to many oil companies is around 15 per cent. In forming our provisions these parameters have been kept closely in mind.

The government's total financial "take" from production of *oil* will be made up from:

   (i) a 1¼ per cent *royalty* on well-head value of production;
  (ii) a *petroleum income tax* at a rate of 50 per cent of taxable income;
 (iii) an *additional profits tax* at a rate of 50 per cent of after tax cash flow after the total investment has been recouped with a money return of 25 per cent;
 (iv) *state participation* in development of commercial fields on the "carried interest" principle as outlined earlier.

The sum effect of these provisions with a 30 per cent state participation is to provide a total state "take" on a medium-sized field of 65-70 per cent of the (discounted) total profits over the field life. The Papua-New Guinea system has the advantage, however, of being considerably more flexible than other systems such as production sharing because the burden of tax is related directly to field profitability. In the case of economically marginal fields the company will retain a greater proportion of the cash flow and the state share will be reduced, while on very profitable fields the state "take" will exceed 70 per cent. The overall effect of the proposed financial package is to make it rather more generous than a 70:30 production-sharing contract (see Tables A and B).

In the case of *natural gas* developments a profits tax is less applicable. Instead there will be a variable charge equivalent to a royalty amounting to 8–16 per cent of the well-head value depending on production level in addition to the petroleum income tax at a rate of 50 per cent.

B. *Oil royalty*

This will be a flat-rate charge of 1¼ per cent of the well-head value of all oil produced. The well-head value of the oil will be calculated by taking the value at the point of export (determined in accordance with the later section on pricing) and deducting allowable costs incurred in transporting the oil from the well head to the point of export. The royalty will be paid in cash and will be an allowable deduction in assessment of petroleum income tax. This royalty is very low in relation to royalties elsewhere and has been designed, in conjunction with the additional profits tax, to ensure that royalty charges do not prevent the development of marginal fields that would be sub-economic with higher royalty rates. The royalties received by the government from on-shore oil production will be paid to the Provinces in which the oil is located.

C. *Petroleum income tax*

The government intends to establish a Petroleum Income Tax Act which will govern the taxation arrangements for all companies in Papua-New Guinea engaged in the exploration, development and production of crude oil and natural gas. The Petroleum Income Tax Act and the additional profits tax and the provisions and rates contained in the Act will apply uniformly to all companies.

The special features of the oil business require separate tax treatment. The separate tax Act will make it quite clear that petroleum production is to be treated separately from other activities. It will also ensure that foreign oil companies can obtain maximum benefit of tax credits in their home countries.

The liquification and transportation of liquified natural gas will be included within the activities covered by the Petroleum Income Tax Act, but refining of petroleum and other "downstream" activities will be taxed under the existing Income Tax Act. The additional profits tax will apply only to producers of oil but not to gas production.

The rate of petroleum income tax will be 50 per cent of taxable income for all petroleum producers, but they will not be subject to dividend withholding tax. The amalgamation of company tax and dividend withholding tax maximises the tax credit benefits for foreign companies, thereby reducing their home taxes and, hopefully, increasing the funds they have available to spend in Papua-New Guinea. This rate of tax is higher than the company tax rate applicable to other companies in Papua-New Guinea. However, petroleum production is a very profitable industry and can easily afford a 50 per cent tax. In very few countries is the rate of company tax on petroleum production less than 50 per cent.

In addition, if the company tax rate is less than 50 per cent in Papua-New Guinea then many oil companies based in other countries and producing oil in Papua-New Guinea would get no benefit from the lower tax rate. They would have to pay tax on Papua-New Guinea income at the lower rate to Papua-New Guinea and then pay extra tax at home at a rate equal to the difference between their home tax rates and Papua-New Guinea tax rates.

The petroleum income tax will be assessed for each company on all income and expenditure from exploration and production within the boundaries of each prospecting licence. This means that only expenditures incurred within the prospecting licence area will be deductible from income earned in the licence area. However, provision will be made for deduction of unsuccessful exploration costs against income. The government considered at length whether both petroleum income tax and additional profits tax should be assessed on a licence-by-licence basis or on a country-wide basis. The licence-by-licence basis of assessment best preserves the intention of taxing oil fields on the basis of profitability. It also means that a company cannot delay paying tax on the profits of one field by offsetting the development costs of another field. This practice is also consistent with government policy in the metal mining industry. The government recognises that a licence-by-licence basis of tax assessment is somewhat more severe on companies than a country-wide basis, and this has been allowed for when considering the overall package of financial measures.

In determining gross income before deductions, the state will value the oil produced on the basis of the norm price system described below. It is intended that the norm price should normally equal the price at which the oil is actually sold, and that this should be a world market price.

The definition of allowable deductions from gross income will be specified in detail in the new Petroleum Income Tax Act. While the treatment of allowable deductions will in most cases be fairly standard, some special provisions will be required. The government recognises that the major participants in the oil industry are huge, multinational corporations who seek to minimise world-wide taxes and, in some cases, have devised measures for shifting untaxed profits from one country to another. Measures will be needed to limit these practices. Some of the proposed allowable deductions are outlined below.

   (i) Current operating costs will be deductible.

  (ii) Operating losses may be carried forward up to 7 years, but not carried back.

 (iii) No deductions will be allowed for technology payments to parent or affiliated companies including royalties and licence fees for the use of inventions and processes. It is not appropriate that the parent company should earn both a "reasonable return" which reflects the service and technology which they provide *and* be paid special fees for those same services. It is also true that elsewhere such payments have been used as a means of transferring profits out of the country tax free.

 (iv) Management fees will be allowable deductions, but the government will retain the right to assure itself that the fees are appropriate.

  (v) Deductions in respect of interest paid on loan funds will be limited to an amount not exceeding an interest rate of 1 per cent above the long-term, prime corporate borrowing rate at the time of borrowing in the lending country. This is a necessary provision to prevent intra-company lending at excessive interest rates which can also be used to reduce tax. As an additional precaution all borrowing will be subject to normal government regulation under the NIDA Act and foreign exchange regulations.

 (vi) Amortisation of exploration expenditures will be on a licence-by-licence basis. All exploration expenditure incurred within the prospecting licence in which a field is located prior to the commencement of commercial production and not more than 11 years prior to issue of the development licence will be eligible to be written off against income earned from within the prospecting licence. Similarly, all exploration expenditure incurred within the prospecting licence after commencement of commercial production will be eligible for write-off commencing in the year of expenditure. In all cases the rate of amortisation will be flexible up to a maximum rate of 20 per cent per annum on a straight line basis. Given the large size of existing prospecting licences and the generous rate of amortisation, these provisions should encourage continued exploration activity. All exploration incurred within the prospecting licence more than 11 years before issue of a development licence will not be deductible. The prospecting licence system envisages 11 years as the maximum length of a licence after which, if no discovery is made, the licence terminates and those expenditures become sunk costs. Applying this logic consistently, exploration costs incurred more than eleven years before a discovery are considered to be sunk costs. These expenses are recovered as part of the very high "reasonable return" which the oil companies obtain from successful fields.

(vii) Exploration expenditures incurred in prospecting licences where no commercial discoveries are made may be amortised against income from successful areas commencing from the time when the prospecting licence in which the unsuccessful exploration was located is fully relinquished. Amortisation will be at a flexible rate of up to 20 per cent per annum on a straight-line basis. This provision will apply to exploration expenditures incurred no more than 11 years prior to relinquishment.

(viii) All capital expenditure incurred in bringing a discovery into production will, in principle, be an allowable capital expenditure, deductible for tax purposes. The new Act will attempt to define clearly which items are or are not deductible. This expenditure will, subject to the next paragraph, be deductible at a rate up to 10 per cent per annum or the life of the asset if that life is less than 10 years.

(ix) Provision will be made for accelerating deductions for capital expenditure and exploration expenditure in years of low cash flow. Accelerated deductions will be allowed in any year up until such time as the sum of after-tax income plus tax depreciation equals the original capital investment, including allowable exploration costs. This period will be known as the investment recovery period. During the investment recovery period additional deductions will be allowed in any year in which after-tax income plus total depreciation equals 25 per cent of the investment, subject to the limitation that additional deductions will not be allowed so as to create a loss for tax purposes, and that total deductions in respect of any item of expenditure or asset shall not exceed 100 per cent of the actual cost of that item. This provision is flexible and provides considerable assistance for projects with low cash flow in the eary years.

(x) Where necessary, regulations will provide for the method of allocation of exploration and capital expenditures between licences.

D. *Additional Profits Tax (APT)*

Certain oil fields may be so productive as to give rise to extremely high profits — well in excess of the expected rate of return required to attract companies to explore. It is well-established Paupa-New Guinea Government policy that, when profits from a natural resource project are higher than this "reasonble return", then the state will take an increasing share of these high profits by way of an additional tax on income. This principle will also be applied to the production of oil (but not to natural gas, for reasons given later). The APT will be applied only after a project has actually achieved a predetermined rate of profitability, measured in terms of a discounted cash flow (DCF) internal rate of return on total funds (both equity and loan funds). After the DCF rate of return has been achieved, the APT will be applied at a uniform rate on after-tax cash flow (net profit plus depreciation less capital expenditure).

The APT proposed here has several important advantages. First, the tax is related directly to the achieved profitability of the venture. It cannot deter development of marginal fields because for a marginal field the APT will never be paid. However, in the case of a very profitable field, the average tax rate adjusts itself automatically to obtain for the state a higher share of the benefits. The APT is a very flexible system and should, therefore, provide much greater assurance of stability over time than any other system. The APT will be assessed on a licence-by-licence basis, but a uniform rate of APT and threshold rate of return will be applicable to all companies producing in Papua New Guinea.

The setting of the rate of APT and the threshold rate of return at which the APT commences are closely linked. And the values of those parameters are clearly critical to the companies in determining the attractiveness of petroleum operations in Papua-New Guinea. Consequently the determination of appropriate values for these parameters has taken up more time than any other aspect of policy formation. Besides extensive consultations with companies and independent consultants we have commissioned computer studies comparing the relative severity of the PNG proposals with those obtaining in other countries where oil exploration, but not production, is under way. Throughout several important facts have guided the deliberations.

First, since no commercial discoveries have been made in Papua-New Guinea, despite fairly extensive past explorations, the risks of failure are high. We recognise that it would not be appropriate to set fiscal terms in Papua-New Guinea equivalent to those applicable in proven oil-producing countries.

Secondly, we recognise that to obtain a high continuing rate of exploration, the financial terms must be sufficiently attractive to induce additional exploration expenditures.

Third, we recognise that in deciding to assess petroleum income tax and additional profits tax on a licence-by-licence basis the required "expected rate of return" will be raised above the level which might be appropriate for a county-wide basis of assessment.

In setting the overall fiscal terms, the government has formed the judgement that, given the present world petroleum situation, a marginal petroleum field would be one that indicated an "expected rate of return" on total funds in the project amounting to about 13–15 per cent in real terms, or about 23–25 per cent in money terms (depending on the expected long-term rate of inflation). Given the usual gearing for such projects the return on shareholders funds would, of course, be very substantially in excess of these returns, and should adequately recompense shareholders for their substantial risks. Thus the threshold rate of return has been set so that APT will not be applicable to a field which just earns over its life a DCF internal rate of return on total funds of 25 per cent. To ameliorate the risk of sustained very high rates of inflation, companies may opt for a variable threshold rate of return equal to 15 per cent plus an agreed inflation factor equal to some agreed rate of interest, in which case the reasonable return would be linked to international interest rates.

The rate of APT will be 50 per cent of the sum of after-tax profit plus depreciation less capital expenditure. The maximum marginal rate of tax on taxable income is thus 75 per cent after the project has achieved a return of 25 per cent. Examples of the way in which this tax system might function for a "typical" Papua New Guinea oil field are set out in Tables A and B.

It should be apparent that even after the threshold rate of return is reached, and APT begins to operate, the company still has substantial potential for increased returns, as it continues to receive 25 per cent of the profits even when the higher tax is being paid. Since the APT will not be applicable until late in the life of the project when depreciation provisions will be largely exhausted then any capital expenditure will reduce the effective rate of tax. In this manner a positive inducement to capital additions, such as secondary recovery, is built into the system.

The calculations involved in the additional profits tax are presented below.

1. *Payments* in any year equal the sum of all payments made in the prospecting licence area for exploration expenditure and capital expenditure, plus all payments allowable as revenue deductions under the

231

Petroleum Income Tax Act, with the exception of interest payments, plus all petroleum income tax actually paid in that year. For purposes of APT all exploration expenditure within the prospecting licence area incurred prior to the commencement of production and not more than 11 years prior to the allocation of the development licence will be treated as payments in the year in which the development licence is issued. Exploration expenditures in unsuccessful areas outside the licence area, incurred not more than 11 years prior to relinquishment, will be treated as payments at the time when the prospecting licence in which the unsuccessful exploration took place is totally relinquished.

2. *Receipts* for any year equal the sum of all proceeds of sales accruing in that year excluding any interest payments receivable. Loan repayments made by the state in payment for its participation share will be receipts for this purpose, but will not be taxable income.

3. *Net Cash Receipts (NCR)* in any year is the excess (positive or negative) of *Receipts* over *Payments* for that year.

4. Determination that a project has achieved the threshold rate of return will be made by the following procedure:
   (a) for each year, an *accumulated value* (AV) of *net cash receipts* (NCR) will be calculated;
   (b) AV for the year in which capital expenditure for development commences will be NCR for that year (which will be negative) including all allowable past exploration expenditure;
   (c) for each subsequent year $AV_{(n+1)} = AV_n(1+a) + NCR_{(n+1)}$ where a = the specified threshold rate of return;
   (d) if at the end of any year AV is positive, this indicates that the threshold rate of return has been achieved and additional profits tax will be charged;
   (e) AV will be deemed to be zero at the end of any year in which additional profits tax is paid, and no further additional profits tax is payable in respect of any year for which AV is negative. This means that if in any year net cash receipts is negative then the project will revert to paying the ordinary petroleum income tax until such time as the cash deficit has been recouped, together with the threshold rate of return.

5. Additional profits tax will be paid at a rate of 50 per cent of net cash receipts.

6. Additional profits tax will be assessed either in kind or in United States dollars, whichever currency is elected by the company concerned at the time of commencement of construction.

232

E. *Other minor taxes*

Oil companies will pay normal import duties and levies, except on equipment imported for exploration purposes and then re-exported. However, it is not intended that import duties on capital goods will create any significant financial burden on a project. At the present time capital goods are not subject to import tariffs and there is payable only a 2½ per cent general import levy. Other minor charges such as local government rates on the unimproved value of land in urban areas, motor vehicle registration fees and normal service charges for services provided by the government will apply to oil companies in a non-discriminatory manner.

F. *Natural gas provisions*

In the case of liquefied natural gas, there will be a variable charge depending on the volume of production which will be financially equivalent to a royalty and will vary from 8–16 per cent of the well-head value of output, depending on the rate of production. However, the charge will not be a royalty but a *free participation*, which will give the state the right to participate on equal terms in decision-making without any cost in the operation. This arrangement will allow greater state involvement in any gas venture without any financial burden on the companies beyond the same level of royalty which they, in any case, would expect to pay in most countries. In addition, LNG producers will be subject to the petroleum income tax and the same depreciation arrangements as oil producers. This arrangement in the case of LNG has been preferred to the APT because of the difficulties in defining profits when so much of the development costs are shipping and de-gasification costs outside of Papua-New Guinea.

In the case of operations producing both oil and gas, or facilities serving both oil and gas fields (port facilities, for example), allowable capital and operating costs will be apportioned between the projects by the Chief Collector of Taxes, in proportion to the facilities' relative use by the two kinds of production (as might be indicated, for example, by the relative values of production for oil and gas).

Pricing and marketing

A. *General problems*

In most industries, it is fair to assume that the actual sales price of a product reflects its value in the market; the sales price is the result of a bargain based on the supply and demand positions of the buyer and the seller – it is an "arm's length" transaction between parties which are both anxious to gain the maximum economic advantage. But this situation does not exist in the petroleum industry. Because of the high degree of concentration in the industry, and the integration, from well to refinery to service station, within the major companies, it is very difficult to determine what a realistic "arm's length" price is for oil products. This is especially true in the case of crude oil, which is usually sold between units of the same international corporation. The most important considerations in the companies' setting of prices between affiliates may not be market conditions, but rather the possible tax advantages of shifting profits from one country of operations to another. In contrast to this position of the companies, it is in Papua-New Guinea's interest to insist on the highest reasonable market price being paid for its products, so as to maximise the taxation revenue and foreign exchange earnings produced by petroleum.

Even sales between un-affiliated companies may not always be at free-market levels. For example, companies may exchange oil supplies in different parts of the world, or link a sales agreement in one country to another sales agreement elsewhere. Prices set in these sorts of agreements may often be arbitrarily low. And, while there are "spot" or free-market sales in oil, these often reflect immediate and localised conditions — haphazard local shortages or surplus — and not a more general supply and demand position. Recently, the producing countries themselves have been selling large amounts of oil directly, using their share of participation oil. But these governmental sales may also fail to reflect free-market conditions. Often oil sales by governments have special prices for political purposes — as in some recent sales by the People's Republic of China — or they may be linked to commitments from foreign industry to construct factories in, or supply capital goods to the oil-producing countries.

The world market for petroleum is thus for a variety of reasons not a free market. This has continued to be true in recent years as governments have stepped in to set the price directly, primarily through OPEC. In this situation the only way in which the state can assure a fair return, both in taxes and other forms of revenue, is to set the value of its oil directly. Most other countries in the world now follow this system, and Papua-New Guinea also intends to follow it.

B. *The norm price*
The government intends to establish a *norm price* (which will, wherever possible, be the free market price) for each petroleum product sold from Papua-New Guinea. This norm price will be used to assess oil producers' taxable income, to compute the value of royalties, and to establish the value of the state's participation share of oil production. The details of

how the norm price will be determined for different classes of product are set out below.

(i) *Natural gas*. Generally the norm price for natural gas will be the export price specified in a sales contract. Normally gas is sold on long-term contracts at specified prices. If this is the case in Papua-New Guinea, the government can assure a fair price through the exercise of its supervisory powers to control sales agreements under the National Investment and Development Act and provisions in the Petroleum Agreements.

If industrial developments within Papua-New Guinea lead to domestic use of natural gas — particularly in industries whose owners are in many cases the same companies that are now involved in natural gas exploration on-shore — the government will need to consider further steps for establishing the value of the gas thus sold.

(ii) *Crude oil*. The norm price will be based on the following standards, and will not be higher than the highest figure derived from them.

(a) The actual realised price in third-party sales of equivalent grade oil produced in the same area;

(b) the price of Light Arabian crude oil, with appropriate adjustments for gravity and quality differences and for differential transport costs; and

(c) the price of Sumatran Light crude oil exported from Indonesia, with appropriate adjustments for gravity and quality and differential transport costs.

In all cases the norm price shall be set from the point of export. While this figure will be different from the well-head value used for calculation of Royalty, it will be a straightforward matter to arrive at a well-head value by deducting from the norm price the costs attributable to operations between the well head and the point of export.

The norm price will be set from time to time by the Minister on advice from the National Petroleum Authority. The norm price will be an average price, applying to production throughout the period, and so will not necessarily reflect the actual sale price of any particular shipment.

The government intends to afford oil companies full opportunity to discuss the norm price before any final determination is made. In most cases it should be possible to reach agreement with the companies on the appropriate price, but the government will have final power to set a price if no agreement is obtained. The price so set by the government will apply for purposes of valuation of oil for royalty, tax and participation purposes.

Even though actual sales by any one producer during a period may be at slightly higher or lower prices than the norm price, this system offers substantial administrative advantages, without being in the long run unfair to any producing field.

NOTES TO TABLES A AND B:

The two cash flow exhibits (set out as Table A and B) illustrate the mechanics of the financial proposals. The fields shown are hypothetical only. The cost figures are based on rule-of-thumb estimates made by petroleum consultants. The onshore field has relatively low exploration and development costs and is highly profitable. The off-shore field has higher costs. All calculations are in real terms – that is, in constant 1975 United States dollars with no allowance for inflation. The company has a 70 per cent share and the state 30 per cent in these cases.

All allowable exploration prior to the first commercial discovery is brought to account in the year of commencement of development. Both on-shore and off-shore fields are rated at 50,000 barrels/day. For the on-shore field first-year production is 50 per cent of peak rate followed by 4 years of peak rate then declining by 10 per cent per annum. For the off-shore field first-year production is 45 per cent of peak rate followed by 4 years at 90 per cent then declining by 10 per cent per annum. Operating costs are assumed constant in each year. Amortisation is at the maximum allowable rates, namely 20 per cent per annum for exploration expenditure and 10 per cent per annum for capital expenditure. Accelerated depreciation in the off-shore case is as described in the text.

Company revenue and costs are 70 per cent of the total reflecting their share. All cash available to government from its 30 per cent participation after meeting costs is used to repay loan account amounts outstanding until the balance is zero. A zero real interest rate is used for simplicity. Tax and royalty money is not used for loan repayment. Calculation of timing of additional profits tax is described in the text. It has been assumed that Petroleum Income Tax is paid in the year it is earned.

The exhibits illustrate two important points. First, in the case of a highly profitable field, such as the on-shore field illustrated, the company earns a return on capital very substantially in excess of the threshold return for additional profits tax. In the on-shore case the company earns a real return of 30 per cent. Secondly, in the less profitable off-shore field the company cash flow is substantially higher than would be the case for a 70:30 production-sharing arrangement with 40 per cent cost recovery oil. The present value of company cash flow discounted at 10 per cent (real) is 30 per cent higher under the PNG system than for the 70:30 production share. This is the case over a wide range of hypothetical fields. In the case of less profitable fields the state share is less than under fiscal terms of other countries while in the case of very profitable fields the state share rises significantly but does not prevent company rates of return achieving extremely generous levels.

Table A

PROFITABLE ONSHORE FIELD 50,000 bb day PRODUCTION

Year	1	2	3	4	5	6	7	8	9	10	11	12	13	14	15	16	17	18	19	20
Project																				
(a) Exploration	10	3.6	8.7	9.3																
(b) Capital Expenditure					50	79.2	4.4													
(c) Operating Cost							7.8	7.8	7.8	7.8	7.8	7.8	7.8	7.8	7.8	7.8	7.8	7.8	7.8	7.8
(d) Output (m. bbs)							9.1	18.2	18.2	18.2	18.2	16.4	14.7	13.3	11.9	10.8	9.7	8.7	7.8	7.1
(e) Value of output							109.2	218.4	218.4	218.4	218.4	196.5	176.4	159.6	142.8	129.6	116.4	104.4	93.6	85.2
Company																				
(f) Sales receipts							76.4	152.9	152.9	152.9	152.9	137.8	123.5	111.7	100.0	90.7	81.5	73.1	65.5	59.6
(g) Costs							5.5	5.5	5.5	5.5	5.5	5.5	5.5	5.5	5.5	5.5	5.5	5.5	5.5	5.5
(h) Depreciation							13.7	13.7	13.7	13.7	13.7	9.3	9.3	9.3	9.3	9.3				
(i) Royalty							1	1.9	1.9	1.9	1.9	1.7	1.5	1.4	1.2	1.1	1	.9	.8	.8
(j) Trading Profit							56.2	131.8	131.8	131.8	131.8	121.3	107.2	95.5	84	74.8	75	66.7	59.2	53.3
(k) Petroleum Income Tax							28.1	65.9	65.9	65.9	65.9	60.6	53.6	47.8	42	37.4	37.5	33.4	29.6	26.6
(l) Additional Profits Tax									9	39.8	39.8	35	31.5	28.5	25.7	23.4	18.8	16.7	14.8	13.4
(m) Govt. Loan Payments							29.2	19.1												
(n) Net Cash Receipts					-81.6	-79.2	+66.6	+98.7	+79.6	79.6	79.6	70	62.9	57	51.3	46.7	37.6	33.3	29.6	26.7
(o) Accumulated Value @ .15					-81.6	-173	-132.4	-53.6	+18											
(p) Cash Flow to Company					-81.6	-79.2	+66.6	+98.7	+70.6	39.8	39.8	35	31.5	28.5	25.7	23.4	18.8	16.7	14.8	13.4
Government																				
(q) Loan A/c Borrowing					24.5	23.8	1.3													
(r) Value of Partn. Oil							32.8	65.5	65.5	65.5	65.5	59	52.9	47.9	42.8	38.9	34.9	31.3	28.1	25.6
(s) Costs							2.3	2.3	2.3	2.3	2.3	2.3	2.3	2.3	2.3	2.3	2.3	2.3	2.3	2.3
(t) Loan Repayment							29.2	19.1												
(u) Loan A/c Balance							19.1	0												
(v) Cash from Partn.							0	0	9	39.8	39.8	35	31.5	28.5	25.7	23.4	18.8	16.7	14.8	13.4
(w) Petroleum Income Tax							28.1	44.1	63.2	63.2	63.2	56.7	50.6	45.6	40.5	36.6	32.6	29	25.8	23.3
(x) Additional Profits Tax								65.9	65.9	65.9	65.9	60.6	53.6	47.8	42	37.4	37.5	33.4	29.6	26.6
(y) Royalty							1	1.9	1.9	1.9	1.9	1.7	1.5	1.4	1.2	1.1	1	.9	.8	.8
(z) Govt. Cash Flow							29.1	111.9	140	170.8	170.8	154	137.2	123.3	109.4	98.5	89.9	80	71	64.1

DCF internal rate of return to company = 30% real terms

Table B[78]
SMALL OFFSHORE FIELD 45,000 bb/Day PNG SYSTEM

Year	1	2	3	4	5	6	7	8	9	10	11	12	13	14	15	16	17	18	19	20
Project																				
Exploration	50																			
Capital Expenditure	1.5	12.7	13.5		46.2	146.3	76.4													
Output							8.2	16.4	16.4	16.4	16.4	14.8	13.3	12	10.8	9.7	8.7	7.5	7.1	6.4
Value of Output							98.4	196.8	196.8	196.8	196.8	177.6	159.6	143.6	129.2	116.3	104.7	94.2	84.8	76.3
Operating Costs							11.8	11.8	11.8	11.8	11.8	11.8	11.8	11.8	11.8	11.8	11.8	11.8	11.8	11.8
Company																				
Sales Receipts							68.9	137.8	137.8	137.8	137.8	124.3	111.7	100.5	90.4	81.4	73.3	65.9	59.4	53.4
Costs							8.3	8.3	8.3	8.3	8.3	8.3	8.3	8.3	8.3	8.3	8.3	8.3	8.3	8.3
Normal Depreciation							29.7	29.7	29.7	29.7	29.7	18.8	18.8	18.8	18.8	18.8				
Accelerated Depreciation							30													
Royalty							.9	1.7	1.7	1.7	1.7	1.6	1.4	1.2	1.1	1	.9	.8	.7	.7
Trading Profit							0	98.1	98.1	98.1	98.1	95.6	83.2	72.2	62.2	53.5	64.1	56.8	50.4	44.4
Petroleum Income Tax							0	49.1	49.1	49.1	49.1	47.8	41.6	36.1	31.1	26.7	32.1	28.4	25.2	22.2
Additional Profits Tax												0.5	30.2	27.5	24.9	22.7	16.0	14.2	12.6	11.1
Government Loan Repay							26													
Cash Flow					-123.9	-146.3	9.3	55.5	101.3	78.8	78.8	66.1	30.2	27.5	24.9	22.7	16.0	14.2	12.6	11.1
Net Cash Receipts					-123.9	-146.3	+9.3	134.2	101.3	78.8	78.8	66.6	60.4	54.9	49.9	45.5	32.1	28.4	25.2	22.2
Accumulated value at 15%					-123.9	-288.8	-322.8	-237.0	-171.2	-118.1	-56.9	+1								
Government																				
Loan A/c Borrowing					37.2	43.9	22.0													
Value Partn. Oil							29.5	59	59	59	59	53.3	47.9	43.1	38.8	34.9	31.4	28.3	25.4	22.9
Costs							3.5	3.5	3.5	3.5	3.5	3.5	3.5	3.5	3.5	3.5	3.5	3.5	3.5	3.5
Loan Repayment							26	55.5	22.5											
Loan Account Balance							78	22.5	0											
Cash from partn.							0	0	33	55.5	55.5	49.8	44.4	39.6	35.3	31.4	27.9	24.8	21.9	19.4
Petroleum Income Tax							0	49.1	49.1	49.1	49.1	47.8	41.6	36.1	31.1	26.7	32.1	28.4	25.2	22.2
Additional Profits Tax												0.5	30.2	27.5	24.9	22.7	16.0	14.2	12.6	11.1
Royalty							.9	1.7	1.7	1.7	1.7	1.6	1.4	1.2	1.1	1	.9	.8	.7	.7
Government Cash Flow							.9	50.8	83.8	106.3	106.3	99.7	117.6	104.4	92.4	81.8	76.9	68.2	59.4	53.4

Correction: Due to computation error, the total depreciation and amortisation deductions exceed the total capital expenditure by US$30 million. This affects cash flow in years 15 and 16 when deductions should be 7.6 and 0 respectively. This does not significantly alter the DCF rate of return.

NOTES

1. R. F. Mikesell, "Financial Considerations in Negotiating Mineral Development Agreements", Paper presented at the Inter-regional Workshop on Negotiation and Drafting of Mining Development Agreements, Buenos Aires, November 1973 (mimeographed), pp. 16-17.
2. Thomas W. Wadde, "Lifting the Veil from Transnational Mineral Contracts: A Review of Recent Literature", *Natural Resources Forum*, 1977, Vol. I, p. 280.
3. *Investment Appraisal*, National Economic Development Office, HMSO, p. 4.
4. Mikesell, op. cit., p. 3.
5. M. Tanzer, *Alaska's Prudhoe Bay Oil: Profitability and Taxation Potential* (A Report to the Alaska State Legislature), 9 January 1976, p. 34.
6. J. Montel, "Concession versus Contract" in Cleland and Seymour (eds), *Continuity and Change in the World Oil Industry*, pp. 111-112.
7. Ibid.
8. F. M. Hooke, "Financial Problems of Exploitation of Petroleum Resources", Paper presented at the Commonwealth Law Conference, Edinburgh, 1977.
9. Mikesell, op. cit., p. 8.
10. Ibid., pp. 7-8.
11. Ibid., p. 7.
12. Ibid., p. 7.
13. Ibid., p. 12.
14. Ibid., p. 13.
15. Tanzer, op. cit., p. 75.
16. Interviews with representatives of oil companies operating in Indonesia.
17. R. F. Newby, "Acquisition of Exploration and Production Rights outside of Canada", *Alberta Law Review*, 1976, Vol. XIV' p. 397.
18. Ibid.
19. E. Penrose, "Profit Sharing between Producing Countries and Oil companies in the Middle East", *The Economic Journal*, 1959, Vol. 69, p. 245.
20. Interviews with representatives of oil companies.
21. R. W. Mikesell, "Conflict in Foreign Investor–Host Country Relations: A Preliminary Analysis", in *Foreign Investment in the Petroleum and Mineral Industries*, p. 36.
22. In interviews with representatives of companies operating in Britain and Norway they frankly expressed their opposition to these taxation measures, and indicated that they had seriously contemplated taking legal action to challenge the validity of action which retroactively changed the terms on which they had been granted exploration rights.
23. *Report* No. 60, 1974-75, Finance Committee, Norwegian Storting (English translation – mimeographed), p. 4.
24. *Official Report*, Parliamentary Debates, House of Commons, 1974-75, Vol. 882, p. 462 – Mr. Edmund Dell, Paymaster-General.
25. *Report* No. 60, op. cit., pp. 5-10.
26. K. W. Dam, *Oil Resources – Who Gets What How?*, p. 126.
27. Ibid.
28. Ibid., p. 127.
29. Summary report entitled "A Comparison of Tax Systems", based on the study made by J. Morgan and Colin Robinson, published in *Petroleum Economist*, 1976, Vol. 43, pp. 170-172.
30. Interviews with representatives of US oil companies (July 1977).
31. *Report on Matters Concerning the Natural Gas Industry in British Columbia*, British Columbia Energy Commission, 1973, Part 4, p. 7.
32. Ibid., pp. 7-8.

33. Ibid., Part 4, p. 9.
34. Ibid., Part 4, p. 13.
35. *Final Report on 1976 Petroleum and Natural Gas Price and Incentives Hearings*, British Columbia Energy Commission, setting out recommendations contained in the interim report, p. 12.
36. Ibid., pp. 56–57.
37. *Statement of Policy on Proposed Petroleum and Natural Gas Act*, Ministry of Energy, Mines and Resources, Canada, 1976, p. 17.
38. M. Faber, "The Fiscal Regime" in *Some Policy and Legal Issues affecting Mining Legislation and Agreements in African Commonwealth Countries*, p. 34.
39. Ibid., pp. 34–35.
40. *Petroleum Economist*, 1976, Vol. 43, p. 482.
41. Faber, op. cit., p. 69.
42. Tanzer, op. cit., pp. 77–78.
43. Ibid., p. 80.
44. Ibid.
45. Dam, op. cit., p. 14.
46. Mikesell, Paper, op. cit., p. 15.
47. D. N. Smith and L. T. Wells, Jr, *Negotiating Third World Mineral Agreements*, pp. 94–95.
48. US Internal Revenue Service Ruling, 1976, published in *Petroleum Legislation: Basic Oil Laws and Concession Contracts: Asia and Australasia*, Supp. 51.
49. John M. Blair, *The Control of Oil*, p. 203.
50. Faber, op. cit., p. 31.
52. A. R. Thompson, "Australian Petroleum Legislation and the Canadian Experience", *Melbourne University Law Review*, 1968, Vol. 6, p. 370 at pp. 390–391.
53. *Norwegian Royal Decree of 8 December 1972 relating to Exploration for and Exploitation of Petroleum in the Sea-bed and Substrata of the Norwegian Continental Shelf*, Section 26.
54. Dam, op. cit., pp. 134–135.
55. D. N. Smith and L. T. Wells, Jr. "Mineral Agreements in Developing Countries", *American Journal of International Law*, 1975, Vol. 69, p. 560 at p. 570.
56. *First Report from the Committee of Public Accounts* (North Sea Oil and Gas), House of Commons, Session, 1972–73, p. 88.
57. Ibid., p. xxii.
58. *White Paper on U.K. Offshore Oil and Gas Policy*, July 1974 (Cmd Paper 5696), paras 14 and 15.
59. Dam, op. cit., pp. 135–136.
60. Ibid., p. 136.
61. F. Tugwell, *The Politics of Oil in Venezuela*, pp. 92–93.
62. R. Engler, *The Politics of Oil*, p. 155.
63. Smith and Wells, op. cit. (Negotiating . . .) p. 73.
64. Smith and Wells, op. cit. ("Mineral Agreements . . ."), p. 587.
65. R. Fabrikant, "Production Sharing Contracts in the Indonesian Petroleum Industry", *Harvard International Law Journal*, 1975, Vol. 16, p. 303 at pp. 339–340.
66. M. Adelman, "Is the Oil Shortage Real?" *Foreign Policy*, 1972, Vol. 9, p. 84.
67. Smith and Wells, op. cit. ("Mineral Agreements . . ."), p. 573.
68. Dam, op. cit., p. 140.
69. Lord Balogh, "Governments and the North Sea", *The Banker*, 1974, Vol. 124, p. 1065 at p. 1066.
70. Dam, op. cit., p. 141.
71. Ibid.
72. A. R. Thompson, op. cit., p. 390.

73. A. R. Thompson, "Australia's Offshore Petroleum Common Code", *University of British Columbia Law Review*, 1970, Vol. 2, p. 1 at p. 21.
74. Ibid.
75. M. M. Olisa, "Comparison of Legislation affecting Foreign Exploitation of Oil and Gas Resources in Oil-Producing Countries", *Alberta Law Review*, 1972, Vol. 10, pp. 493–505.
76. Ibid., pp. 501–502.
77. The material contained in this Appendix is drawn from a report prepared by a leading firm of petroleum consultants, Walter J. Levy, with their permission, which is gratefully acknowledged. The material has been included to illustrate the methodology used to prepare a comparative evaluation of financial terms under different types of licensing arrangements. Since such evaluations are made as the basis of a variety of assumptions, including those as to price, cost and tax rates, and since in the present case considerable time has elapsed since the report was prepared, the actual figures and computations used should not be taken to be currently applicable.
78. In Table B, due to computation error, the total depreciation and amortisation deductions exceed the total capital expenditure by US $80 million. This affects cash flow in years 15 and 16 when deductions should be 7.6 and 0 respectively. This does not significantly alter the DCF rate of return.

CHAPTER VI

Concluding Observations: Petroleum Development in the Eighties — Prospects and Options for Developing Countries

PRESENT SITUATION AND PROSPECTS FOR PETROLEUM DEVELOPMENT IN DEVELOPING COUNTRIES

With the significant exception of the OPEC countries and a few others, developing countries are oil importers. Of the non-OPEC developing countries only four are major oil exporters — Oman, Brunei, Bahrain and Trinidad and Tobago — and nine are minor oil exporters — Angola, Bolivia, Congo, Egypt, Malaysia, Mexico, Syria, Tunisia and Zaïre, with Peru a potential addition to the latter group. While the market share of the nine minor oil exporters was about 2 per cent of total oil exports, according to World Bank estimates,[1] this could increase to about 7 per cent by 1980 and 8 per cent in 1985.

Past patterns of oil consumption in the developing countries show steady growth, between 1960 to 1975, of about 7.2 per cent per annum. It is estimated that if the developing countries are to sustain a growth rate of about 5.8 per cent between 1975 to 1985, their energy consumption must grow to 5 per cent per annum up to 1980 and to about 7.1 per cent thereafter. Table A sets out the energy balance of oil-importing developing countries, 1975–1985. In order to meet this urgent need, in developing countries, where possibilities exist of tapping domestic resources, high priority is attached to petroleum exploration, since a significant discovery would free them from imports which may well prove to be beyond their capacity to sustain.

An assessment of petroleum prospects in the oil-importing developing countries (OIDCs) made by BEICIP, an affiliate of the French Petroleum Institute, for the World Bank, estimates the "ultimately recoverable oil reserves" (URR) of OIDCs at 60 billion barrels, that is, six times their current proven reserves. While this may form a small percentage of the total global URR (about 4 per cent), if such reserves could become available to the OIDCs they could substantially meet their consumption needs over the next 15 years. The rate of exploration of prospective areas in OIDCs is markedly below that in the rest of the world — thus, the rate of

Table A
PRELIMINARY PROJECTIONS OF OIL-IMPORTING DEVELOPING COUNTRIES' (NODCs) ENERGY BALANCE 1975-85*
(million barrels per day of oil equivalent)

Oil-importing developing countries (OIDCs)		1975	1980	1985	*Growth rates (% per annum)*		
					1976-80	1981-85	1976-85
Consumption:	Oil	4.33	5.35	7.20	4.3	6.2	5.2
	Non-oil	3.73	4.95	7.30	5.8	8.1	6.9
	Total	8.06	10.30	14.50	5.0	7.1	6.0
Production:	Oil	1.21	1.66	2.85	6.5	11.4	8.9
	Non-oil	3.62	4.88	7.35	6.2	8.5	7.3
	Total	4.83	6.54	10.20	6.2	9.3	7.8
Net energy imports		3.23	3.76	4.30	3.3	2.7	3.0
Oil imports		3.12	3.69	4.35	3.4	3.4	3.4
Oil imports as % of total imports		14.4	12.6	7.2			
Value of oil imports (current $ billion/year)		14.3	24.3	38.3			

*Refers to commercial energy sources only and assumes OPEC crude oil prices remain constant in real terms through 1985 ($11.50 per barrel in 1975). OIDCs are projected to grow at: 5 per cent per annum in 1976-80; 6.4 per cent per annum in 1981-85; 5.8 per cent per annum for the whole of the decade.

Source: World Bank staff estimates.

243

proven to ultimate reserves is 17 per cent when that of the rest of the world's is 40 per cent. It is estimated that if present trends were to continue, by 1985 only 20 per cent of the URR of OIDCs will have been discovered, while the rest of the world will have discovered 70 per cent. There are extensive prospects in the developing countries; as many as fifty to sixty developing countries could be expected to join the ranks of the oil-producing by 1985.

In order to realise their potential, however, the overall exploration and development programme in the OIDCs needs to be substantially expanded. A programme, embodying targets regarded as the maximum which, in BEICIP's judgment, could reasonably be developed, was one which would have as its aim the raising of the rates of proven to ultimate reserves to 59 per cent instead of 30 per cent, that could be attained on the basis of present investment trends. The annual production to be aimed at for 1985 would be 256 million metric tons, compared to 52 million metric tons in 1975. This would require a *doubling of the rate of discovery and development and far wider geographical diversification of investments in the OIDCs.*

The results of the successful implementation of such a programme would be significant, so that assuming a 5 per cent growth rate in oil consumption, the oil deficit of the OIDCs would be reduced from 70 per cent of consumption to 6 per cent in 10 years. The OIDCs as a group could expect to become self-sufficient, and thus their annual import bill could be reduced by about $8.6 billion. Most of the larger countries could become nearly self-sufficient, while a good number with small domestic requirements could expect to become exporters, so that total oil exports from the new producers could reach up to about 50–60 million metric tons a year by 1985.

Global petroleum development programme

Requirement of capital, technology and managerial skills
To implement such a programme globally, the requisite capital, technology and managerial skills have to be brought together to undertake extensive programmes of exploration and development in the OIDCs. Capital is required both for exploration and development. The estimated overall capital requirement up to 1985 is of the order of $60 billion (1975 dollars) — of this about half would represent downstream investment and of the other half about $25 billion would be required for production, and $5 billion would be the "risk capital" needed for exploration.

Sophisticated technology as well as managerial skills would also be required to implement the programme.

Role of multinational oil companies

Traditionally, multinational oil companies had provided capital, and technology and managerial skills for petroleum development under concessionary arrangements, which were seen as being particularly advantageous to them. With changes in the global environment, governments have been able to attract capital and technology of the oil companies for petroleum development on more advantageous terms than was the case under traditional concessions. The new types of petroleum arrangements, examined in Chapter IV, show how the governments have steadily been able to improve from their point of view the terms on which petroleum exploration and development rights were to be granted. The "government take" had also tended to increase steadily under the new types of financial arrangements, discussed in Chapter V.

Changing legal relations between governments and multinational oil companies: improved terms under new types of agreements

Under the new petroleum development arrangements considered in Chapter IV a basic element that was discernible was the growing trend towards hybrid systems where general legislation defined the broad policy parameters and laid down minimum standards within which individual agreements were to be negotiated. The new generation of petroleum development arrangements, in contrast to the traditional concessions, had developed various mechanisms to safeguard the interests of the state as the owner of the natural resource. Among the features which distinguished the new arrangements from the traditional concessions, the provisions dealing with the following matters are noteworthy:

(a) Areas to which they applied were limited and clearly defined.
(b) Rights granted were for a limited period.
(c) Separate provisions were made for the exploration phase and the development/production phase.
(d) Minimum work programmes and minimum expenditure obligations were incorporated; carefully formulated minimum work obligation clauses stipulated the nature of surveys to be carried out, the number of wells to be drilled within a stipulated period, and even the depth to which the wells were to be drilled.
(e) Progressive relinquishment of the area during the exploration phase, so that if at the end of the exploration phase, no discovery was made,

then the entire area was to be surrendered.

(f) Specific obligations were laid down in relation to the development and production phase. Thus, an obligation to develop was spelt out, in the event of a commercially significant discovery being made; criteria were laid down for determining whether a discovery was to be regarded as commercially significant or not; directives were contained on the basis of which the rate of extraction/depletion and production methods could be regulated.

(g) Training of nationals at every level and in every phase of the operations.

(h) Procurement of goods and services from domestic sources, supply of oil to the domestic market (often at a concessional price), investment in downstream installations and facilities, and extending of other benefits to the national economy.

(i) Participation by the government or national oil company in the management of operations.

In addition to these provisions which were principally aimed to extend control over operations, the financial provisions under the new arrangements were so designed as to give the government a substantially larger take, and also to appropriate a substantial portion of any "windfall" profits which may be generated. The variety of mechanisms which have been used to achieve these objectives include: levy of tax (with provision for variation of the tax rates), institution of sliding-scale rates for royalties and/or production shares, fiscal provisions so designed that a major part of profits above a certain level, computed in terms of "a rate of return" to the investor, could be appropriated by the state by way of an "excess profits" tax. Other mechanisms included control over pricing or the use of a "norm" or "tax reference" price for computation of tax or other forms of government take. Mechanisms have also been designed for more effective monitoring of costs and for limiting capital and other allowances which could be claimed by companies.

Thus, a government which determines that it would like multinational oil companies to undertake a role in petroleum exploration and development in its territory can, and indeed should, devise arrangements which incorporate provisions and mechanisms to safeguard the public interest. The elements derived from the new generation of agreements which could go into the making of a "good" petroleum development arrangement have been described thus:[2]

. . . it is possible to detect a new direction in many of the recent oil exploration agreements between the international oil companies and developing countries all over the world. The main consideration in these agreements has been the quick development of resources with minimum cost to the host countries. To meet this objective, precautions have to be taken while negotiating with the foreign investor regarding the length of the exploration period, obligations of investment and work, continuous revisions [sic] of work done, obligations to hand over to the government any information obtained, participation and training of local personnel in the exploration work, recovery costs by the foreign investor during the period of exploration only in the event of commercial discovery of oil, manner of operation of joint ventures, fixation of an adequate rate of return to the foreign investor, and finally the over-all duration of the contract. For example, it would be possible to fix the time limit for exploration to about three years. During this period, the terms should specify clearly the amount of obligatory investment and the schedule of work to be done by the foreign investor, including the number of wells to be drilled and so forth. In the light of the progress made in the initial three year period, an extension may be granted for an additional period to account for uncertainties involved (possibly complicated by shortage of materials and fluctuating world prices). Six years should provide more than enough time for any company either to find oil or to decide to give up. All explorations, development risks, and expenses should be the responsibility of foreign investors. In the event of successful discovery and possibilities of establishing commercial production a joint company with majority government participation could be formed and a rate of return mutually agreed, perhaps linked with production levels. The over-all duration of the contract, including the production period, could be designed for short periods ranging from five to ten years, with provision for extension and revision of the terms at intervals.

Most of the features of a sound petroleum development arrangement are contained in the above enumeration. A model agreement for a particular country would, however, have to be designed keeping in view on the one hand, its own priorities and objectives, and on the other, the "attractiveness" of its territory to potential investors in terms of geological prospects, history of earlier explorations, stable operating environment, etc. For while it would be contrary to the national interest to grant petroleum exploration and development rights to companies on the basis of an arrangement that did not adequately safeguard the public interest, it would be counter-productive to devise a "model agreement" which contained provisions that were totally unacceptable to potential investors. The essential point is to know where to draw the line, or to strike the balance — so that the capital, technology and managerial skills of the companies can be engaged on the best possible terms. The capacity to exercise fine judgment in the matter would require an understanding of the global environment — the world petroleum industry and the petroleum market — as well as information about the company's background and its corporate strategy and policies derived from its record of past and current operations, and also a sound estimate of the geological prospects of the territory involved. Government negotiators may be expected to come closer to striking the right balance the greater their level of understanding and information of the matters indicated. They must be sensitive as to what is regarded as important or relevant by the other side in order to be able to negotiate effectively. Where lack of knowledge or experience exists, as is likely to be the case with many government negotiators, it would be prudent to draw upon professional expertise and counsel.

The question, however, remains that while the role that multinationals can play in petroleum development is recognised, and governments may be willing, and some may indeed be keen to attract them, there is evidence, discussed in Chapter II, to indicate that multinational oil companies are not themselves disposed to undertake a significantly *expanded* programme of exploration in the developing countries.

To achieve a significant expansion of petroleum development programmes in the developing countries — or to approach the target suggested above of doubling the present rate of exploration — concerted action is needed, not only at the national, but also at the regional and international level, along the lines indicated in Chapter III.

Action at the national level

At the national level governments can pursue thier objectives of accelerated petroleum development by themselves arranging for more extensive geological and geophysical surveys to be carried out with technical and financial assistance from international agencies or under bilateral inter-governmental arrangements. There may even be a case for some exploratory drilling being undertaken by the governments, utilising the services of specialised contractors. The elaboration of a legal and policy framework (or where one has been in existence for some time, its reappraisal), and the setting up of administrative machinery, manned by technically qualified and competent personnel, merit high priority. The capabilities of the personnel can be strengthened by training and by provision of the "right" kind of technical assistance, that is, the kind which would help to develop their own capabilities and progressively reduce their dependence on externally procured technical assistance.

Action at the regional level

Regional co-operation could materially contribute towards the expansion of petroleum development programmes in the developing countries. The areas for regional co-operation which were identified by the Economic Commission for Africa, and which are just as relevant for other regions, include:

(a) Co-operation among members in the search for new oil/gas deposits.
(b) Periodic meetings of experts to exchange information on the petroleum industry.
(c) Preparation of feasibility studies for sub-regional projects.
(d) Co-ordinated plan for the establishment of refineries in the region.
(e) Survey of all existing training and research facilities and manpower requirements in the field of petroleum and the establishment of new centres and institutions.
(f) Strengthening of existing training centres and institutions.
(g) Establishment of a regional petroleum institute.
(h) Exchange of petroleum information and visits of experts.
(i) Formation of joint field prospecting parties to undertake inventories of oil resources.
(j) Joint ventures for exploration of oilfields common to more than one country and exchange of exploration data.
(k) Co-operation in the exploitation of common fields astride national boundaries by drawing up common plans for exploitation, exchanging exploitation data and establishing common production policies.
(l) Establishment of a regional documentation centre, for the collection, analysis and dissemination, on a continuing basis, of up-to-date information on all aspects of the petroleum industry.
(m) Establishment of a regional petroleum development agency/unit to provide advisory or promotional services.

To the above could be added the establishment of regional service or operating companies, which could draw upon technical resources available within a region. These companies could, as specialised contractors, undertake such operations as geological and geophysical surveys, and exploratory drilling.

Action at the international level

Some of the types of action identified as appropriate for regional co-operation could with equal effect be taken within a framework of international, inter-governmental co-operation. This could be done within the

United Nations framework, and indeed within that of the Commonwealth. For within the Commonwealth, there exist valuable opportunities for a pooling of resources, drawing upon the capital, technology and managerial skills of some of its members, which would make possible effective programmes of co-operation, be it in the form of exploration operations, joint ventures, or provision of technical and managerial services in the field of petroleum development.

The need for new initiatives at the international level has been receiving increasing recognition. The Bonn Summit Declaration (17 July 1978) urged that "the World Bank explore ways in which its activities in this field can be made increasingly responsive to the needs of the developing countries, and . . . examine whether new approaches, particularly to financing hydrocarbon exploration, would be useful". The UN Secretary-General endorsed the conclusions of a group of experts, appointed by him, which had recommended, thus:

> Consideration should be given to expansion of the existing activities of the World Bank and the regional Development Banks, and in particular to the provision of loans for basic geological survey and geo-scientific data base activities . . .
>
> Urgent consideration should be given to additional financing mechanisms and, in particular, a mechanism for providing further finance for petroleum exploration in the oil importing developing nations — this to be linked to the work currently underway within the World Bank and other forums.

Existing international organisations can play a more active role in promoting petroleum development. They can act as a "catalyst" to bring together from different sources in the public and the private sector, financial, technological and managerial resources needed to support and expand exploration programmes in developing countries. They can help to put together new packages of terms under which petroleum development ventures could be promoted. The new policy orientation of the World Bank in this field envisages a more active role for the Bank, thus:[4]

> The trend, therefore, in Bank activities in the sector is an upward one. But questions remain: how fast and to what extent can the Bank expand its operations? And what should the Bank's main role be in helping finance projects in the sector? . . . the Bank can hope to finance only a small portion of the financial needs of the sector . . . Most of the funds will have to come from foreign investors and the producing nations, international financial institutions such as the Bank would need to contribute only so much as to permit them to assume the responsibility for appraising and supervising projects and

helping producing nations marshal the total funds required, including co-financing with other external agencies. Thus, in this sector, too, the Bank's major role would be a catalytic one . . . because financing of exploration activities is an area best left to risk capital, the Bank's role in this area would be a minor one, at the margins, and only in special cases; for geological, geochemical, and geophysical reconnaissance, and for better appraisal of promising discoveries.

The Bank's participation in projects could certainly have the effect of inducing a more positive response from the oil companies. Home governments (in consumer countries) of the companies could also develop arrangements which could induce a larger response from companies. Thus, in the field of hard minerals, proposals have been made by mining companies for a more active role by the European Economic Community which call for provision of financial support or guarantees, the setting up of an insurance fund to cover the risk of political action by host governments, and the conclusion of co-operation agreements between home governments and host governments, under which home governments can take up matters where companies complain of violations of their legal rights. It is considered that a more feasible initiative would be a scheme for collective financing by EEC of exploration, the *quid pro quo* being an undertaking by host governments to assure supplies to the EEC members financing the scheme.[5]

Given the inhibition on the part of multinational oil companies to a major expansion of exploration operations in developing countries, considerable importance must be attached to initiatives which can be taken by international organisations to support state-owned oil entities and public sector organisations to make a larger contribution to petroleum exploration. Such efforts would merit strong support from consumer governments.

Three specific initiatives at the international level which could give the required push towards a significant expansion of petroleum exploration in developing countries are:

Establishment of a financial facility for petroleum exploration
A financial facility such as a revolving fund could provide necessary "risk capital" for exploration in areas where capital may not readily be available from other sources. Since the magnitude of the capital required for exploration is small compared to that required for development, a revolving fund of even $500 million could provide a significant stimulus to the rate of exploration. The fund could be reimbursed in case a commercial discovery was made in the course of an exploration financed by the fund.

Inventory of technological and managerial resources

A global stock-taking of existing unutilised or under-utilised capacity in the field of petroleum exploration would be useful. The object of the exercise would be to locate all available sources which developing countries could draw upon to procure technical and managerial services required to undertake different aspects of exploration. Such services could be procured on the basis of service contracts.

Establishment of international machinery to promote petroleum exploration in developing countries

Some international machinery could be established to act as an institutional entrepreneur, or "catalyst", to identify and promote feasible petroleum exploration and development ventures in developing countries. The type of machinery envisaged is not another large bureaucracy, but effective cells which could be developed in existing international organisations. Such machinery could provide valuable assistance in promoting petroleum development ventures by bringing together the financial, technological and managerial skills necessary to establish such ventures under new types of petroleum development arrangements.

NOTES

1. The statistical information and estimates set out in this section are based on World Bank sources, and on a study by BEICIP, an affiliate of the French Petroleum Institute, which was commissioned by the Bank.
2. R. Vedavalli, *Private Foreign Investment and Economic Development – A Case Study of Petroleum in India*, pp. 197–198.
3. *Multilateral Development Assistance for the Exploration of Natural Resources*, Report by the Secretary-General of the United Nations, A/33/256, 16 October 1978, incorporating the report of the group of experts.
4. World Bank, *Annual Report*, 1978, p. 22.
5. M. Faber and Roland Brown, *Changing the Rules of the Game* (Report of the Workshop on Mining Legislation and Mineral Resources Agreements organised by the Commonwealth Secretariat and the UN Centre on Transnational Corporations, October 1978), esp. at p. 2 and pp. 36–37.

Table B

OIL-IMPORTING DEVELOPING COUNTRIES' ENERGY BALANCE 1960–85
(assuming medium projection of GNP growth and $11.50 price (in 1975 $))
(million b/d of oil equivalent))

	1960	1970	1973	1974	1975	1976	1977	1980	1985
Inland Consumption									
Oil	2.8	5.5	6.8	7.0	7.2	7.8	8.4	9.8	12.5
	(1.5)	(3.3)	(4.2)	(4.3)	(4.3)	(4.4)	(4.5)	(4.8)	(5.4)
Non-Oil	(1.3)	(2.2)	(2.6)	(2.7)	(2.9)	(3.4)	(3.9)	(5.0)	(7.1)
Production									
Oil	1.8	3.3	3.6	3.8	4.0	4.6	5.2	6.7	9.5
	(0.5)	(1.1)	(1.1)	(1.1)	(1.1)	(1.1)	(1.3)	(1.5)	(2.3)
Non-Oil	(1.3)	(2.2)	(2.5)	(2.7)	(2.9)	(3.5)	(3.9)	(5.2)	(7.2)
Bunkers (all oil)	0.2	0.3	0.3	0.3	0.3	0.3	0.3	0.4	0.5
Net Imports	1.2	2.6	3.5	3.5	3.5	3.5	3.5	3.5	3.5

Source: 1960–74 World Energy Supplies, 1950–74.
UN Series J. 19, United Nations.
1975–85, World Bank Estimates (Economic Analysis and Projections Department).

Bibliography

Books

ADELMAN, M. A., *Alaskan Oil*, New York: Praeger, 1971.
ADELMAN, M. A., *The World Petroleum Market*. Baltimore: Johns Hopkins University Press, 1972.
AL-OTAIBA, M. S., *OPEC and the Petroleum Industry*. London: Croom Helm, 1975.
BALL, D. K. and TURNER, D. S., *This Fascinating Oil Business*. New York: The Bobbs-Merill Co. Inc., 1965.
BALLEM, J. B., *The Oil and Gas Lease in Canada*. Toronto: University of Toronto Press, 1973.
BARNET, R. J. and MULLER, R. E., *Global Reach*. New York: Simon Schuster, 1974.
BARROWS, Gordon H., *The International Petroleum Industry*. New York: International Petroleum Institute, 1965.
BARTLETT, A. G. III, *Pertamina, Indonesia National Oil*. Djakarta: Amerasian, 1972.
BERMUDEZ, A. J., *The Mexican National Petroleum Industry*. A Case Study in Nationalisation. Stanford: Stanford University Press, 1963.
BLAIR, J. M., *The Control of Oil*. London and New York: Macmillan Press, 1977.
BRADLEY, P. G., *The Economics of Crude Petroleum Production*, Amsterdam: North Holland Publishing Co., 1967.
BROWN, E. H., *The Saudi Arabia-Kuwait Neutral Zone*. Beirut: Middle East Research and Publishing Centre, 1963.
BROWN, R. and FABER, M., *Some Policy and Legal Issues affecting Mining Legislation and Agreements in African Countries*. London: Commonwealth Secretariat, 1977.
CAMPBELL, Robert W., *The Economics of Soviet Oil and Gas*. Baltimore: Johns Hopkins University Press, 1971.
CATTAN, H., *The Evolution of Oil Concessions in the Middle East and North Africa*. New York: Oceana, 1967.
CATTAN, H., *The Law of Oil Concessions in the Middle East and North Africa*. Dobbs Ferry, New York: Oceana, 1967.
CONNELLY, P. and PERLMAN, R., *The Politics of Scarcity* (Resource Conflicts in International Relations). London: Oxford University Press, 1975.
DAM, Kenneth W., *Oil Resources: Who Gets What How?*. Chicago and London: University of Chicago Press, 1976.
DASGUPTA, B., *The Oil Industry in India*. London: Frank Cass, 1971.
DECHERT, C. R., *Ente Nazionale Idrocarburi – Profile of a State Corporation*. Leiden: E. J. Brill, 1963.
DE CHAZEAU, M. G. and KAHN, A. E., *Integration and Competition in the Petroleum Industry*. New Haven: Yale University Press, 1959.
DEMARIS, Ovid, *Dirty Business*. New York: Harper's Magazine Press, 1974.
ENGLER, R., *The Politics of Oil*. New York: Macmillan, 1961.

ERICKSON, Edward W. and WAVERMAN, Leonard (eds). *The Energy Question: An International Failure of Policy, Vol. 1. The World and Vol. 2 North America.* Toronto: University of Toronto Press, 1974.

FABRIKANT, R., *Oil Discovery and Technical Change in South-East Asia* (Legal Aspects of Production Sharing Contracts in the Indonesian Petroleum Industry). Singapore: Institute of South-East Asian Studies, 1972.

FABRIKANT, R. (collected by), *The Indonesian Petroleum Industry: Miscellaneous Source Materials.* Singapore: Institute of South-East Asian Studies, 1973.

FINNIE, D. H. *Desert Enterprise: The Middle East Oil Industry in its Local Environment.* Cambridge, Mass: Harvard University Press, 1958.

FIRST, R., *Libya: The Elusive Revolution.* Middlesex: Penguin Books, 1974.

FRANKEL, P. H., *Essentials of Petroleum, a Key to Oil Economics.* London: Frank Cass, 1975.

FRANKEL, P. H., *Mattei: Oil and Power Politics.* London: Faber and Faber, 1966.

FRIEDMANN, W., K. and BEGUIN, J. P., *Joint International Business Ventures in Developing Countries.* New York: Columbia University Press, 1971.

GOELLER, Harold E., *World Energy Conference Survey of Energy Resources.* New York: U.S. National Committee of the World Energy Conference, 1974.

HARTSHORN, J. E., *Oil Companies and Governments. An Account of the Industrial Oil Industry in its Salient Environment.* London: Faber and Faber, 1962.

HOWELL, L. E. and MORROW, M., *Asia Oil Politics and the Energy Crisis.* IDOC/ International Documentation Nos. 60–61, 1974.

HIRST, D., *Oil and Public Opinion in the Middle East.* New York: Praeger, 1966.

ISSAWI, C., *Oil, the Middle East and the World.* London/Beverly Hills, California: Sage Publications, 1972.

ISSAWI, C. and YEGANEM, M., *Economics of Middle Eastern Oil.* New York: Praeger, 1962.

JACOBY, N. H., *Multinational Oil.* New York: Macmillan, 1974.

KRUEGER, Robert A., *The United States and International Oil: A Report for the Federal Energy Administration on U.S. Firm and Government Policy.* New York: Praeger, 1974.

KUBBAH, A. A. E., *Libya: its Oil Industry and Economic System.* Baghdad: Arab Petroleum Economic Research Centre, 1964.

LAN, Quah Swee, *Oil Discovery and Technical Change in South-East Asia: A Preliminary Bibliography.* Singapore: Institute of South-East Asian Studies, 1971.

LENCZOWSKI, G., *Oil and State in the Middle East.* Ithaca: Cornell University Press, 1960.

LIEUWEN, Edwin, *Petroleum in Venezuela: A History.* Berkeley: University of California, 1954.

LONGRIGG, Stephen, *Oil in the Middle East: Its Discovery and Development.* London: Oxford University Press, 1968.

LOVEJOY, W. F. and HOMAN, P. T., *Economic Aspects of Oil Conservation Regulations.* Baltimore: Johns Hopkins University Press, 1967.

LUTFI, A., *OPEC Oil.* Beirut: Middle East Research and Publishing Centre, 1968.

MACGREGOR HUTCHESON, A. and HOGG, A., *Scotland and Oil.* Edinburgh: Oliver and Boyd, 1975.

MACKAY, D. I. and MACKAY, G. A., *The Political Economy of North Sea Oil.* London: Martin Robertson, 1975.

MCLEAN, J. G. and HAIGH, R. E., *The Growth of Integrated Oil Companies.* Cambridge: Harvard University Press, 1954.

MADELIN, H., *Oil and Politics.* Farnborough: Saxon House/Lexington Books, 1973.

MARTINEZ, A. R., *Chronology of Venezuelan Oil.* London, 1969.

MIKDASHI, Z., *A Financial Analysis of Middle Eastern Oil Concessions 1901–65.* New York: Oceana, 1966.

MIKDASHI, Z., *The Community of Oil-Exporting Countries.* London: George Allen and Unwin, 1972.

MIKDASHI, Z., *International Politics of Natural Resources*. Cornell: Cornell University Press: 1977.

MIKDASHI, Z., CLELAND, S. and SEYMOUR, I., *Continuity and Change in the World Oil Industry*. Beirut: Middle East Research and Publication Centre, 1970.

MIKESELL, R. F., *Non-Fuel Minerals, U.S. Investment Policies Abroad*. London: Sage Publications, 1975.

MIKESELL, R. F., *Foreign Investment in the Petroleum and Mineral Industries*. Baltimore: Johns Hopkins Press, 1971.

MIKESELL, R. F. and CHENERY, H. B., *Arabian Oil*. Chapel Hill: University of North Carolina Press, 1949.

MORAN, T. H. *Multinational Corporations and the Politics of Dependence*. Princeton: Princeton University Press, 1974.

MUGHRABY, M. A., *Permanent Sovereignty over Oil Resources*. Beirut: Middle East Research and Publishing Centre, 1966.

ODELL, P. R., *Oil and World Power*. Middlesex: Penguin Books, 1970.

PAYNE, B. and ZORN, J. (eds), *Foreign Investment, International Law and National Investment*. Sydney: Butterworth, 1975.

PEARSON, Scott, R., *Petroleum and the Nigerian Economy*. Stanford: Stanford University Press, 1970.

PEARTON, M., *Oil and the Romanian State*. Oxford: Clarendon Press, 1971.

PENROSE, Edith., *The Large International Firm in Developing Countries: The Petroleum Industry*. London: George Allen and Unwin, 1968.

POSNER, M. V. and WOOLF, J. S., *Italian Public Enterprise*. London: Gerald Duckworth & Co., 1967.

RAJARATNAM, M., *Politics of Oil in the Philippines*. Singapore: Institute of South-East Asian Studies, 1973.

ROSTOW, E. V., *A National Policy for the Oil Industry*. New Haven: Yale University Press, 1948.

POWELL, J. R., *The Mexican Petroleum Industry*. New York: Russel & Russel, 1956.

SAMPSON, A., *The Seven Sisters (The Great Oil Companies and the World they Make)*. London: Hodder and Stoughton, 1975.

SAYEGH, K. S., *Oil and Arab Regional Development*. New York: 1968.

SELL, G. (ed.), *Competitive Aspects of Oil Operations*. London: Institute of Petroleum, 1958.

SHWADRAN, B., *The Middle-East, Oil and the Great Powers*. New York: Praeger, 1959.

SMITH, D. N. and WELLS, L. T., Jr, *Negotiating Third World Mineral Agreements*. Cambridge, Mass.: Ballinger Publishing Co., 1975.

STEVENS, P. J., *Joint Ventures in Middle-East Oil 1957–1972*. Unpublished thesis, University of London.

STOCKING, G. W., *Middle-East Oil*. London: Allen Lane (Penguin Press), 1970.

SYMONDS, E., *Oil Prospects and Profits in the Eastern Hemisphere*. New York: First National City Bank, 1962.

TANZER, M., *The Political Economy of International Oil and the Under-Developed Countries*. Boston: Beacon Press, 1969.

TANZER, M., *The Energy Crisis*. New York and London: Monthly Review Press, 1974.

TORIGUIAN, Shavarsh, *Legal Aspects of Oil Concessions in the Middle East*. Beirut: Hameskaim Press, 1972.

TUGENDHAT, C. and HAMILTON, A., *Oil, the Biggest Business*. London: Eyre & Methuen, 1975.

TUGWELL, F., *The Politics of Oil in Venezuela*. Stanford: Stanford University Press, 1976.

TURNER, L., *Oil Companies in the International System.* London: Allen and Unwin, 1978.

VEDAVALLI, R., *Private Foreign Investment and Economic Development – A Case Study of Petroleum in India.* Cambridge: Cambridge University Press, 1976.

VOTAW, D., *The Six Legged Dog, Mattei and ENI: A Study in Power.* Berkeley: University of California Press, 1966.

WIDSTRAND, W., *Multinationals in Africa.* Stockholm: Almqvist and Wikesell, 1975.

WILLIAMS, H. R. and MEYERS, C. J., *Manual of Oil and Gas Terms.* New York: Matthew Bender, 1971.

WILLRICH, M., *Energy and World Politics.* New York: The Free Press, 1975.

Articles

ADELMAN, M., "Is the Oil Shortage Real?", *Foreign Policy*, 1972, p. 69.

ADELMAN, M. A., "Oil Price in the Long Run (1963–75)", *The Journal of Business*, University of Chicago, April 1964.

AMERASINGHE, C., "Ceylon Oil Explorations", *American Journal of International Law* (USA), April 1964, Vol. 58, p. 445.

BALFOUR, R. J., "Basic Provisions Governing Duration of a Lease", *Alberta Law Review*, 1965–66, Vol. 4, p. 219.

BALOGH, Lord, "Government and the North Sea", *The Banker*, September 1974.

BALOGH, Lord, "The Scandal of the Great North Sea Give-Away", *Sunday Times* (London), 13 February 1973.

BEAUCHAMP, K., CROMMELIN, M. and THOMPSON, A. R., "Jurisdictional Problems in Canada's Offshore", *Alberta Law Review*, 1973, Vol. II, p. 431.

BELL, R. D., "Taxation of Mining and Petroleum", *Alberta Law Review*, 1974, Vol. 12, p. 36.

BEN BELLA, Ahmed, "Algeria's Oil Policy", *Review of Contemporary Law*, 1964, No. 2, p. 16.

CALKINS, H. H., "Drilling Contract – Legal and Practical Considerations", *Rocky Mountain Mineral Law Institute*, 1976, Vol. 21, pp. 285–306.

CARLSTON, K. S., "International Role of Concession Agreements", *North Western University Law Journal*, 1957–58, Vol. 52, p. 618.

CARLSTON, Kenneth S., "International Role of Concession Agreements", *North Western University Journal*, 1957–58, Vol. 52 (p. 618).

CHANDLER, Geoffrey, "The Myth of Oil Power – International Groups and National Sovereignty", *International Affairs*, 1970, Vol. 46, No. 4.

CLARKSON, Kenneth W., "The Economics of Work Requirements in Ocean Resources", *Virginia Journal of International Law* (1975), Vol. 15, p. 795.

COLITTI, M., "Vertical Integration, Multinational Oil Companies and Newcomers", Paper read at the Petroleum Economics Seminar, University of Oxford, March 1976.

CROMMELIN, M., "Queensland Oil and Gas Law: Comments", *Queensland Law Journal*, 1970–2, Vol. 7, p. 292.

CROMMELIN, M., "Jurisdiction over Onshore Oil and Gas in Canada", *University of British Columbia Law Review*, 1975, Vol. 10, No. 1.

CROMMELIN, M., "Offshore Oil and Gas Rights", *Natural Resources Journal*, October 1974, Vol. 14, pp. 457–500.

CURRAN, J. F., "Effect of Amendments to Petroleum and Natural Gas Leases", *Alberta Law Review*, 1965–66, Vol. 4, p. 267.

DAKIN, A. N., "Future Patterns of Legislation in the Petroleum Industry", *University of Melbourne Law Review*, 1967, Vol. 6, p. 68.

DAM, K. W., "Oil and Gas Licensing and the North Sea", *Journal of Law and Economics*, Vol. 8, 1965, p. 51.

257

DAM, K. W., "The Evolution of North Sea Licensing Policy in Britain and Norway", *Journal of Law and Economics*, 1974, Vol. 17, p. 213.

DAM, K. W., "The Pricing of North Sea Gas in Britain", *Journal of Law and Economics*, 1970, Vol. 13, p. 11.

DAVIES, G. J., "Legal Characterization of Overriding Royalty", *Alberta Law Review*, 1972, Vol. 10, p. 232.

ELY, Northcutt, "Policy Considerations in the Development of Mineral Laws", *Natural Resources Lawyer*, 1970, Vol. III, No. 2, p. 281.

ELY, Northcutt, "The Conservation of Oil", *Harvard Law Review* (1938), Vol. 51, pp. 1209-44.

ELY, N. and PIETROWSKY, R. F., "Changing Concepts in the World's Mineral and Petroleum Development Laws", *Brigham Young University Law Review*, 1976, pp. 9-35.

FABRIKANT, R., "Production Sharing Contracts in the Indonesian Petroleum Industry", *Harvard International Law Journal*, 1975, Vol. 16, p. 303.

FAVRE, J., "Exploration in Petroleum Deficit Countries", Paper read at UN Meeting on Petroleum Cooperation among Developing Countries, Geneva 1975 (ESA/NRET/AC. 10/19).

FIORINA, D. J., "Judicial-Administrative Interaction in Regulatory Policy-Making: The Case of the Federal Power Commission", *Administratie Law Review* – Administrative Law Section – American Bar Association, 1976, Vol. 28, pp. 41-88.

FISHER, M. J. and GOLBERT, A., "*British Petroleum* v. *Libya*: A Preliminary Comparative Analysis of the International Oil Companies Response to Nationalisation", *South Western University Law Review*, 1975, Vol. 7, pp. 68-97.

FLEMING, S., "The Oil Companies prepare for the day when the oil runs out", *Financial Times*, 24 May 1976, p. 14.

FRANKO, L. G., "Arab Countries and Western Oil Companies: Is Cooperation possible?", Paper read at the Seminar on Administration of Oil Resources of Arab Countries (sponsored by the Arab Institute for Social and Economic Planning, Kuwait), Tripoli, April 1974.

FRICK, Robert, H., Comments. Proceedings of the 67th General Annual Meeting 1973 American Society of International Law, 1973, Vol. 67.

GESS, K. N., "UN Resolutions on the Permanent Sovereignty Over Natural Resources", *International and Comparative Law Quarterly* 1964, Vol. 13, pp. 398-450.

GIBSON, J., "Production Sharing", *2 Bulletin of Indon. Econ. Studies* (February 1966), p. 52 and (June 1966), p. 75.

GOLDIE, L. F. E., "The Exploitability Test – Interpretation and Potentialities", *1968 Natural Resources Journal* (July), pp. 434-477.

GRAHAM, S. B., "Fair Share – or Fair Game? Great Principle, Good Technology – But Pitfalls in Practice", *Natural Resources Lawyer*, 1975, Vol. 8, pp. 61-88.

GRAY, A. D., "North Sea: Foreign Tax Planning for Oil and Gas Producing and Service Operations", *Tulane Tax Institute*, 1975, Vol. 24, pp. 354-71.

GROSSLING, B. F., "Latin America's Petroleum Prospects in the Energy Crisis", *US Geological Survey*, Vol. 141.

GROSSMAN, R. and JOHNSON, C. R., "Distinction Between Exploration and Development Expenditure in the Hard Minerals Industry", *Tax Lawyer* (Taxation Section of the American Bar Association), Fall 1973, Vol. 27, pp. 119-147.

GUILDBERG, T., "International Concessions: A Problem of International Economic Law", *15 Nordisk Tiddskrift for Int. Ret.*, 1944, p. 47.

HAMILTON, R. E. and KRONING, L. J., "Federal-Provincial Tensions in Canada's Emerging Oil Policy", *Law and Policy in International Business*, Vol. 7, 1975, p. 173.

HODGSON, R. D. and ALEXANDER, L. M., "Towards an Objective Concept of Special Circumstances", Occasional Paper No. 13, Law of the Sea Institute (University College, Law Library, London).

HOLLOMAN, J. H., Jr, "Are Overriding Royalties Unrelated Business Income?", *Oil and Gas Tax Quarterly*, Spring 1975, Vol. 24, pp. 1–12.

HOOKE, F. M., "Financial Problems of Exploitation of Petroleum Resources", Paper presented at Commonwealth Law Conference, Edinburgh, 1977.

HUNTER, A., "Oil Developments", *7 Bulletin of Indonesian Econ. Studies*, March 1971, p. 98.

HYDE, J. N., "Economic Development Agreements", *105 Hague Receuil*, 1962, Vol. I, p. 272.

HYDE, J. N., "Permanent Sovereignty Over Natural Resources and Wealth", *American Journal of International Law*, Vol. 50, 1956, pp. 854–867.

JACKSON, H. M., "Rational Development of Outer Continental Shelf Oil and Gas", *Oregon Law Review*, 1975, Vol. 54, pp. 567–581.

JENNINGS, R. Y., "The Limits of Continental Shelf Jurisdiction", *International and Comparative Law Quarterly*, 1969, Vol. 18, p. 8.

KEOHANE, Robert O. and OOMS, Van Doorn, "The Multinational Firm and International Regulation", *International Organization*, Vol. 29, No. 1, p. 169.

KILLEY, J. M., "Drilling and Service Contracts in Offshore Oil and Gas Operations", *Alberta Law Review*, 1973, Vol. 11, p. 480.

LEVY, W. J., "Basic Considerations for Oil Policies in Developing Countries in Techniques of Petroleum Development", *Proceedings of the United Nations International Seminar on Techniques of Petroleum Development*. January–February 1962, New York: U.N. 1964.

LINDEN, W. M., "Oil and Gas Depletion Regulations – Complexity Compounded", *Oil and Gas Tax Quarterly*, March 1976, pp. 351–383.

LOGIE, J., "Les Contrats Petroliers Iraniens", *Revue Belge de Droit International*, 1965, pp. 392–428.

LOUMIET, C., "Toward an International Commodity Agreement on Petroleum", *Denver Journal of International Law and Politics*, 1975, Vol. 5, pp. 485–523.

LUBELL, Harold, "The Soviet Oil Offensive", *Quarterly Review of Economics and Business*, November 1961, p. 11.

MCDONALD, J. G., "Current Developments in Oil and Gas Income Taxation", *Alberta Law Review*, 1976, Vol. 14, No. 3.

MACKAY, J. R., "Considerations in the Search and Exploration for Minerals in British Columbia", *Alberta Law Review*, 1973, p. 538.

MCWILLIAMS, D. A. and MUIR, R. C., "Offshore Operating Agreements", *Alberta Law Review*, Vol. 11, 1973, p. 503.

MIKDASHI, Zuhayr, "Co-operation among Oil Exporting Countries with Special Reference to Arab Countries: A Political Economy Analysis", *International Organization*, 1974, Vol. 28, No. 1, p. 1.

MIRVAHABI, F., "Claims to the Oil Resources in the Persian Gulf: Will the World Economy be Controlled by the Gulf in the Future?", *Texas International Law Journal*, 1976, Vol. 11, pp. 75–112.

MOFFETT, Denny, J., "Federal Oil Shale Policy: An Analysis of Development Alternatives", *Houston Law Review*, 1976, Vol. 13, p. 701 at pp. 719–721.

MORAN, T. H., "Transnational Strategies of Protection and Defence by Multinational Corporations: Spreading the Risk and Raising Cost of Nationalisation in Natural Resources", *International Organisation*, 1973, Vol. 27, No. 2, p. 273.

MUIR, J. D., "Changing Legal Framework of International Energy Management", *International Lawyer*, Fall 1975, Vol. 8, pp. 605–614.

NEWBY, Robert F., "Acquisition of Exploration and Production Rights Outside of Canada", *Alberta Law Review*, 1976, Vol. 14, p. 396.

OLISA, M. M., "Comparison of Legislation affecting Foreign Exploitation of Oil and

Gas Resources in Oil Producing Countries", *Alberta Law Review*, Vol. 10, 1972, p. 487.

ONORATO, W. T., "Apportionment of an International Petroleum Deposit", *International and Comparative Law Quarterly*, 1968, Vol. 17, p. 85.

PEREIRA, Osmy Duarte, "Brazilian Oil Policy and some Legal Aspects of Petrobras, the State-Owned Oil Enterprise", *Review of Contemporary Law*, 1964, No. 1, p. 108.

PENROSE, E. T., "Government Partnership in the Major Concessions of the Middle East", *MEES Supplement*, 30 August 1968.

PENROSE, E. T., "Profit Sharing Between Producing Countries and Oil Companies in the Middle East", *The Economic Journal*, June 1959.

PENROSE, E. T., "Middle East Oil: The International Distribution of Profits and Income Taxes", *Economica*, August 1960.

PENROSE, E. T., "Vertical Integration with Joint Control of Raw Material Production", *Journal of Development Studies*, April 1965.

SMITH, David N. and WELLS, Louis T., "Mining Resources of the Third World: From Concession Agreements to Service Contracts", 1973 American Society of International Law *Proceedings of the 67th Annual General Meeting*.

STONE, Oliver L., "United States Legislation Relating to the Continental Shelf" (with a special reference to the exploitation of oil and gas resources), *International and Comparative Law Quarterly*, 1968, Vol. 17, p. 103.

THOMPSON, A. R., "Australia's Offshore Petroleum Code, *University of British Columbia Law Review*, 1970, Vol. 2.

THOMPSON, A. R., "Australian Petroleum Legislation and the Canadian Experience", *Melbourne University Law Review*, 1968, Vol. 6, p. 370 at pp. 396-391.

THOMPSON, A. R., "Sovereignty and Natural Resources – A Study of Canadian Petroleum Resources", *Valparaiso University Law Review*, 1967, pp. 284-319.

UTTON, A. E., "Institutional Arrangements for Developing North Sea Oil and Gas", *Virginia Journal of International Law*, 1968-69, Vol. 9, pp. 66-81.

VAGTS, D., The Multinational Enterprise: A New Challenge for Transitional Law, *Harvard Law Review*, 1970, Vol. 83, pp. 739-792.

WAELDE, T., "Lifting the Veil from Transnational Mineral Contracts: A Review of the Recent Literature", *Natural Resources Forum*, 1977, Vol. 1.

WAKEFIELD, S. A., "Allocation, Price Control and the FEA: Regulatory Policy and Practice in the Political Arena", *Rocky Mountain Mineral Law Institute*, 1976, Vol. 21, pp. 257-284.

WALDEN, Jerrold L., "The International Petroleum Cartel in Iran – Private Power and Public Interest", *Journal of Public Law*, Spring 1962, Vol. 2, No. 1, p. 17.

WELTER, R. J., "How to Best Achieve the Five Best Tax Objectives in Structuring an Oil and Gas Drilling Deal", *Journal of Taxation*, July 1974, Vol. 41, pp. 38-45.

WILLIAMS, H. R., "Comments on Oil and Gas Jurisprudence in Canada and the United States", *Alberta Law Review*, 1965-66, Vol. 4, p. 189.

WILLIAMS, H. R., "Some Ingredients of a National Gas and Oil Policy", *Stanford Law Review*, Fall 1975, Vol. 27, pp. 969-984.

YAMANI, A. Z., "Oil Industry in Transition", *Natural Resources Lawyer*, 1975, Vol. 8, pp. 3911198.

ZAKARIYA, H. S., "Convergence and Divergence between the Exporting Countries: The OPEC Experience", Paper presented at the International Colloquium on Petroleum Economics at the University of Laval, Quebec.

ZAKARIYA, H. S., "New Directions in the Search for and Development of Petroleum Resources in Developing Countries", *Vanderbilt Journal of Transnational Law*, 1976, Vol. 9, p. 545.

ZAKARIYA, H. S., "Sovereignty, State Participation and the Need to Re-structure the Existing Petroleum Concession", *Alberta Law Review*, Vol. 10, 1972, p. 232.

Official publications and reports

NORWEGIAN, ROYAL: MINISTRY OF FINANCE, Parliamentary Report, No. 25, 1973-74.

NORWEGIAN STORTING: REPORT TO THE NORWEGIAN STORTING
1. No. 30 – (1973-74) Operations on the Norwegian Continental Shelf.
2. No. 60 (1974-75) English Translation.
3. No. 91: On Petroleum Exploration north of 62°N (1975-76).

OAPEC ANNUAL REPORT, 1975.

OECD, Oil: The Present Situation and Future Prospects, Paris: OECD, 1973.

OECD, Energy Prospects to 1985: An Assessment of Long Term Energy Developments and Research Policies (2 vols.) Paris: OECD, 1974.

OFFICIAL REPORT, PARLIAMENTARY DEBATES, House of Commons, 1974-75, Vol. 891.

SELECT COMMITTEE ON NATIONALISED INDUSTRIES – Nationalised Industries and the Exploitation of North Sea Oil and Gas. First Report: 1974-75.

SELECTED DOCUMENTS OF THE INTERNATIONAL PETROLEUM INDUSTRY (OPEC Publication).

UNITED KINGDOM OFFSHORE OIL AND GAS POLICY, CMND Paper 5696, 1974.

UNITED KINGDOM HOUSE OF COMMONS FIRST REPORT of the Committee on Public Accounts on North Sea Oil and Gas, 1972-73.

UNITED NATIONS, Environmental Aspects of Natural Resources Management, New York: U.N. A/Conf. 48/7/Jan. 26, 1972.

UNITED NATIONS, Petroleum Co-operation among Developing Countries, STA/ESA/57, 1975.

UNITED NATIONS YEARBOOK, 1965.

UNITED STATES SENATE, Committee on Foreign Relations, Multinational Petroleum Companies and U.S. Foreign Policy, Washington: Government Printing Office, 1975.

UNITED STATES SENATE, The International Petroleum Cartel, Washington: U.S. Government Printing Office, 1952.

List of Selected Published Petroleum Agreements

S. No.	*Agreement*	*Source**
		PL/ME/BOL/CC
1	Petroleum Concession Agreement between the Ruler of Abu Dhabi and Phillips Petroleum Company, American Independent Oil Company, and AGIP S.p.A., dated 21 January 1967	Supp. XXIII/ OPEC Sel. Doc. 1967 p. 165
2	Off-shore Concession Agreement and Operating Agreement between the Government of Abu Dhabi and Maruzen Oil Co. Ltd, Daikyo Oil Co. Ltd, and Nippon Mining Co. Ltd, dated 6 December 1967	Supp. XXIV/ OPEC Sel. Doc. 1967 p. 137
3	Agreement between Abu Dhabi Government and Pan Ocean Oil Corporation	Supp. XXXIV
4	Agreement between the Ruler of Abu Dhabi and Mitsubishi Mining Co. Ltd, dated 14 May 1968	OPEC. Sel. Doc. 1967 p. 45
5	ADPC (Abu Dhabi Petroleum Company) Agreement, dated 1 June 1970, for Amortisation of Exploration and Drilling Expenditure	Supp. XXXXIV
6	Concession Contract of 15 December 1970 between Superior Oil (Bahrain) Inc. and the Ruler of Bahrain	Supp. XXXII
7	AGIP Mineraria and NIOC Joint Venture Agreement	D-1 Vol. 1
8	Sapphire Petroleum Ltd – and NIOC Joint Venture Agreement	E-1 Vol. 1
9	Joint Structure Agreement between NIOC and Farsi Petroleum (BRP, RAP, SNPA), dated 19 January 1965	Supp. XI
10	Joint Structure Agreement between NIOC and Atlantic Refining Co., Murphy Oil, Sun Oil, and Union Oil, dated 18 January 1965	Supp. VI
11	Joint Structure Agreement between NIOC and Bataafse Petroleum Maatschappij N.V., including related letters and Statutes, dated 16 January 1965	Supp. IX
12	Joint Structure Agreement between NIOC and Tidewater Skelly, Sunray DX, Superior, Kerr-McGee, Cities Service, and Richfield companies for off-shore District I operations, dated 16 January 1965	Supp. XIII
13	ERAP/SOFIRAN Contract for Exploration and Development with NIOC, effective 12 December 1966	Supp. XV

S. No.	Agreement	Source
14	Joint Structure Agreement of 1965 between NIOC and PEGUPCO (operating company for German group)	Supp. XVI
15	Joint Structure Agreement between NIOC, IMINOCO (AGIP S.p.A., Phillips Petroleum Co., and Oil and Natural Gas Commission, India-IMINOCO), dated 17 January 1965	Supp. XVIII
16	Second Supplemental Agreement of 1 June 1973 between the NIOC and AGIP S.p.A.	OPEC/Sel. Doc. IPI 1973 p. 37
17	Exploration and Production Contracting Agreement between NIOC and Continental Oil Co., signed 6 April 1969	Supp. XXVI
18	Text of Joint Structure Agreement between Amerada Hess Corporation and NIOC for Petroleum Exploration and Development (BUSHCO), dated 27 July 1971	Supp. 41/1
19	Service Contract for Exploration and Development of Petroleum between NIOC and Deutsche Erdoelversorgungsgesellschaft m.b. H-Deminex (DEMINEX), dated 30 July 1974	Supp. 46/22
20	Iraq Petroleum Company's Concession Agreement (Concession Agreement (Convention of 14 March 1925, as revised by Principal Agreement of 24 March 1931	OPEC/Sel. Doc. IPI-Kuwait and Iraq pre-1966 p. 7
21	Basrah Petroleum Company's Concession Agreement of 29 July 1938	OPEC/Sel. Doc. IPI-Kuwait and Iraq pre-1966 p. 50
22	Agreement of February 1952 between the Government of Iraq and Iraq Petroleum Company Limited, Mosul Limited, and Basrah Petroleum Company Limited	p. 71
23	Memorandum of Agreement concerning the Border Value of Kirkuk Crude between the Government of Iraq and IPC, MPC, and BPC, dated 24 March 1955	p. 85
24	Working Memorandum relating to the Agreement, dated 3 February 1952, between the Government of Iraq and IPC, MPC, and BPC, dated 24 March 1955	p. 87
25	Memorandum of Agreement between the Government of Iraq and IPC, MPC, and BPC, dated 15 April 1957	p. 93
26	INOC-ERAP Contract in Iraq between Iraq National Oil Company and Enterprise de Recherches et d'Activités Petrolières, dated 3 February 1968	OPEC/Sel. Doc. IPI 1968 p. 107
27	Iraq-IPC Agreement	p. 154
28	Edwin W. Pauley Concession	PL/BOL/ME/CC A-1 Vol. II
29	George Ismiri Concession of 28 November 1957	B-1 Vol. II
30	Jordanian-Yugoslav Oil Concession Agreement (Law No. 23 of 8 March 1968)	Supp. XXVII OPEC/Sel. Doc. IPI 1968 p. 335

264

S. No.	Agreement	Source
31	Shell Group Off-shore Concession of 15 January 1961	Supp. III/OPEC Sel. Doc. Kuwait and Iraq pre-1966 p. 199
32	Concession Agreement between Hispanica de Petroleos S.A. (HISPANOIL) and Kuwait National Petroleum Co. (KNPC) dated 3 May 1967	PL/ME/BOL/CC Supp. XXV/ OPEC Sel. Doc. IPI 1968 p. 155
33	Concession Agreement with Kuwait Oil Company Ltd of 23 December 1934	OPEC/Sel. Doc. IPI-Kuwait and Iraq pre-1966 p. 116
34	Concession Agreement with American Independent Oil Company (Aminoil) of 28 June 1948	p. 127
35	Agreement regarding Crude Oil and Products Dealings with D'Arcy Kuwait Co. and Gulf Kuwait Co. of 11 October 1955	p. 139
36	Agreement regarding Cash Payments with D'Arcy Kuwait Co. and Gulf Kuwait Co.	p. 143
37	Consolidating Supplemental Agreement of 11 October 1955 with D'Arcy Kuwait Co. and Gulf Kuwait Co.	p. 146
38	The Agreement of the New Oil Concession over the Neutral Zone Off-shore Area of 5 July 1958	p. 153
39	Agreement with D'Arcy Kuwait Co. and Gulf Kuwait Co. regarding Selling Charges of 18 December 1958	p. 183
40	Memorandum of Agreement between the State of Kuwait and Kuwait Shell Development Company Limited	p. 227
41	Supplemental Agreement between HH the Ruler of Kuwait and American Independent Company of 29 July 1961	OPEC/Sel. Doc. IPI-Kuwait and Iraq pre-1966 p. 243
42	Supplemental Agreement with BP Kuwait Ltd and Gulf Kuwait Co. Ltd regarding Relinquishment of 17 January 1963	p. 259
43	American Independent Oil Co. (Aminoil) Concession Agreement of 28 June 1948 over Kuwait half of the Neutral Zone	PL/ME/BOL/CC A-1 Vol. II
44	Arabian Oil Co. Ltd Off-shore Concession Agreement of July 1958	B-1 Vol. II
45	Commercial Japanese Petroleum Co. Off-shore Concession Agreement with the Kingdom of Saudi Arabia	D-1 Vol. II
46	Petroleum Agreement between the Sultanate of Oman and the Elf-Sumitomo Group, dated 14 May 1975	Supp. 47 p. 5
47	Qatar Oil Company Ltd (Japan) Concession Agreement, dated 20 March 1969, covering Off-shore Area	Supp. XXIII A-O

266

S. No.	Agreement	Source
	hydrocarbons, approved by Decree No. 61-1045 of 16 September 1961	
63	Sonatrach-Getty Agreement between the Government of Algeria and Getty Oil, dated 19 October 1968	OPEC/Sel. Doc. IPI 1968 p. 253
64	Algeria-France Oil Accord, dated 29 July 1965, between the Governments of France and Algeria	I.L.M. 1965 Vol. 4 p. 809
65	Concession Contract between Pan American UAR Oil Co. and the Government of the UAR, dated 23 October 1963	PL/NA/BOL/CC Supp. II A–1
66	1974 Concession Agreement between Deminex and EGPC	Supp. 28 p. 8
67	Concession Contract between International Egyptian Oil Company, Cie. Orientale des Petroles d'Egypte, and the Government of the UAR, dated September 1963 (ENI/ AGIP Contract)	PL/NA/BOL/CC Supp. III A–1
68	Concession Contract between Pan American UAR Oil Co., Egyptian General Petroleum Corp., and the Government of the UAR, covering the Gulf of Suez Area, dated February 1964	Supp. III B–1
69	Concession Contract, dated September 1963, between Phillips Petroleum Co., the Egyptian General Petroleum Corp., and the Government of the UAR	Supp. IV A–1
70	Petroleum Concession Agreement of 30 September 1969 between the Government of the UAR and the Egyptian General Petroleum Corp. and Pan American UAR Oil Co., covering the Nile Basin area	Supp. XIII A–0
71	Petroleum Agreement of 1971 between the Government of the Arab Republic of Egypt and the Egyptian General Petroleum Corp. and Transworld Petroleum Corp. (this contract never ratified; for final ratified version see Supp. 22)	Supp. XVIII A–0
72	Text of Off-shore Concession Agreement for Exploration and Production of Petroleum between the Government of the Arab Republic of Egypt and Egyptian General Petroleum Corp. and Transworld Petroleum Corp., dated 10 March 1973	Supp. XXII A–0
73	Concession Agreement for Petroleum Exploration and Production between the Government of the Arab Republic of Egypt and the Egyptian General Petroleum Corporation and Petrobras International S.A. Bras-Petro, effective 26 August 1973	PL/NA/BOL/CC Supp. XXIII A–0
74	Concession Agreement for Petroleum Exploration and Production between Arab Republic of Egypt and the Egyptian General Petroleum Corporation and Mobil Exploration Egypt Inc., dated 10 May 1973	Supp. XXIV B–0
75	Concession Agreement for Petroleum Exploration and Production between Arab Republic of Egypt and the Egyptian General Petroleum Corporation and Mobil Exploration Egypt Inc., signed 21 July 1974	Supp. 26 p. 1

S. No.	Agreement	Source
76	Text of Esso Concession Agreement for Petroleum Exploration and Production between the Government of the Arab Republic of Egypt and the Egyptian General Petroleum Corporation and Esso Egypt Inc., dated 22 November 1973	Supp. 27 p. 1
77	Concession Agreement for Petroleum Exploration and Production between the Arab Republic of Egypt, the Egyptian General Petroleum Corporation, and Shell Winning N.V., dated 18 December 1974	Supp. 29 p. 1
78	Concession Agreement for Petroleum Exploration and Production between Egyptian General Petroleum Corporation and Esso, dated 14 December 1974	I.L.M. March 1975 Vol. 14 p. 915
79	Agreement of 1 June between Middle East American Oil Co. and Libyan American Oil Co.	PL/NA/BOL/CC Supp. VI B-0
80	Joint Venture Agreement between AUXERAP (Societe Auxiliare de l'Entreprise de Recherches et d'Activités Petrolières) and AQUITAINE-LIBYE on the one part, and LIPETCO (Libyan-General Petroleum Corp.) on the other, dated 30 April 1968	OPEC/Sel. Doc. IPI 1968 p. 197 or PL/NA/BOL/CC Supp. X A-0
81	Joint Venture Agreement between Libyan General Petroleum Corp. and Ashland Libyan Company, entered into as of 10 May, 1969	Supp. XVII B-0
82	Agreement on Exploration and Production between Libyan National Oil Corp. and Occidental of Libya, Inc., dated 7 February 1974	I.L.M. March 1975 Vol. 14 p. 645
83	Agreement for the Exploration and Exploitation of Hydrocarbons between Bureau de Recherches et de Participations Minières and Maroc Sun Oil Company, dated 10 October 1974	PL/NA/BOL/CC Supp. 29 p. 31
84	Documents pertaining to the Exploration and Exploitation Rights held by Husky Oil Co., including (1) Law 6232 of 25 June 1962, granting Husky Oil Co. the right to Provisions promulgated by Decree of 13 December 1948, with (2) Convention authorising Exploration and Exploitation for Mineral Substances of the Second Group, and (3) Memorandum of Obligations Annexed to the Agreement	Supp. IV A-1
85	Text of Petroleum Concession granted Transworld Tunisia Petroleum Corporation, 1972	Supp. XX A-0
86	Clark Brunei Corporation Petroleum Mining Agreement 1964	PL/A & A/BOL/ CC Supp. III B-1
87	Petroleum Mining Agreement between the State of Brunei and Ashland Oil Company on On-shore State Lands, dated 12 November 1968	Supp. XXXIV A-0
88	The Burma Concession Rules, 1962	Supp. I A-1
89	1970 Myanma Oil Corporation Order as contained in Notification No. 1/70, dated 28 February 1970	PL/A & A/BOL/ CC Supp. XXXI A-0

S. No.	Agreement	Source
90	General Terms and Conditions for Off-shore Oil and Gas Exploration bids as issued by Myanma Oil Corporation, 14 May 1973	Supp. XXXIX A-0
91	Contract of August 1966, between Indonesian Government's Permina and Independent Indonesian American Petroleum Company	Supp. XIII A-0
92	Contract of 6 October 1966, between Indonesian Government's Permina and Japan Petroleum Exploration Co. (JAPEX)	Supp. XIII B-0
93	Agreement for Sale of Assets, between P.T. Shell Indonesia and Indonesian Government, dated 30 December 1965	Supp. XIV A-0
95	Production-Sharing Contract between P.N. Pertambangan Minjak Nasional (Permina) and Phillips Petroleum Company, 1968	OPEC/Sel. Doc. IPI 1968 p. 81 *or* PL/A & A/BOL/ CC Supp. XXIII A-0
95	Production-Sharing Contract between Prof. Dr Wendell Phillips and P.N. Pertambangan Minjak Dan Gas Bumi Nasional (Pertamina), signed 4 February 1970	Supp. XXVI A-0
96	Production-Sharing Contract between Perusahaan Pertambangan Minjak Dan Gas Bumi Negara (Pertamina) and Mobil Petroleum Indonesia Inc., dated 14 March 1973 (Makassar Strait)	Supp. 44 p. 1
97	Production-Sharing Contract between Pertamina and Calasiatic/Topco (Mountain Front-Kuantan), signed 20 January 1975	Supp. 46 p. 1
98	Production-Sharing Contract between Perusahaan Pertambangan Minyak Dan Gas Bumi Negara and Phillips Petroleum Company Indonesia and Tenneco Indonesia Inc., dated 22 March 1975 with amendments (re basic $5 per barrel price), dated 30 June 1975	Supp. 47 p. 7
99	Agreement for Co-operation between Kuwait National Petroleum Company (KNPC) and P.N. Pertambangan Minjak Dan Gas Bumi Nasional (Pertamina), dated 9 September 1968	OPEC/Sel. Doc. IPI 1968 p. 105
100	Agreement for Petroleum Exploration and Production in the Submarine Exploitation Area of Korea between Government of the Republic of Korea and Wendell Phillips Oil Co. (Korea) Ltd, dated 24 December 1970	PL/A & A/BOL/ CC Supp. XXXIII A-0
101	Model Oil Concession Agreement (including annexes) issued by the Government of Pakistan, 1969. Annexes including: (1) Map and Description of Area; (2) List of Machinery and Equipment required; (3) Exploration License Model Deed; (4) Bond Form for Exploration Licence	Supp. XLI A-0

S. No.	Agreement	Source
102	Model Service Contract formulated by Economic and Development Authority, Petroleum Board, Pursuant to the Oil Exploration and Development Act of 1972	Supp. XXXVIII A-0
103	Service Contract between Texaco Philippines, Inc., Chevron Oil Company of the Philippines, Astro Mineral & Oil Corporation, and Jabpract Mining & Industrial Corporation, dated 30 September 1972	Supp. XXXIX A-0
104	Service Contract, dated 23 January 1974, between Philippine Sun Oil Company, Westrans Petroleum Inc., Basic Petroleum and Minerals, Inc., Liberty Mines, Inc., and the Petroleum Board	Supp. 42 p. 46
105	Contract for Exploration and Exploitation of Off-shore Areas between Chinese Petroleum Corporation and Continental Oil Co. of Taiwan, dated 27 March 1971	Supp. XXXIII A-0
106	Triton Oil & Gas Corp. Concession No. 8/2515/12 covering Off-shore Exploration Blocks 18 and 19, signed 12 October 1972	PL/A & A/BOL/ CC Supp. A-0
107	Model Concession Agreement, as published by the Ministry of Economy, Office of the National Petroleum Board, 1973 (Model Concession for Off-shore)	Supp. XXXVII A-0
108	Agreement between Forest Cyprus, dated 10 August 1962	PL/EU/BOL/CC Supp. III A-1
109	Oil Exploration Licence awarded during 1970 to Oxoco Petroleum Corporation	Supp. XVIII A-0
110	Concession Agreement between A. P. Møller and Government of Denmark, as contained in Order 270 of 16 July 1962, by Ministry of Public Works	Supp. II A-1
111	A. P. Møller Concession as modified by Order of 7 November 1963 (this contains the text as originally granted and published in Supplement III, plus the modifications enacted 7 November 1963 to extend it to the Continental Shelf)	Supp. IV A-1
112	Off-shore Concession granted to Texaco Production Services Ltd, dated 28 September 1968	Supp. XV A-0
113	Hydrocarbon Concession Agreement between the Greek State and Ada Oil Exploration Corp., dated 19 December 1969	Supp. XXVII A-0
114	Agreement, dated March 1970, between the Kingdom of Greece and An-Car Oil Company Inc. for exploration and Development on and off-shore Zakynthos, Kyllini, and Cefalonia Islands	Supp. 30 p. 2
115	Greece-Texaco Agreement, dated 28 September 1968	OPEC/Sel. Doc. IPI 1968 p. 289
116	Draft Concession to Explore and Exploit Petroleum in Greenland, prepared April 1974	PL/EU/ BOL/CC Supp. 31 p. 1

S. No.	Agreement	Source
117	Agreement between Government of Ireland and Ambassador Irish Oil Ltd., dated 13 January 1959	Vol. I A–1
118	Government Notice L.n. 41 of 1973 containing terms and Conditions for Off-shore bidding on Sixteen Blocks between dates of 2 May 1973 and 2 August 1973	Supp. XXVI B–0
119	Off-shore Concession Contract, dated 31 August 1973, between Portuguese Government and Portugal Sun Oil/ Amerada Hess of Portugal and Phillips Petroleum Company, Portugal, concerning Dourada and Moreia (Areas 11 and 12)	Supp. 33 p. 13
120	Contract, dated 29 March 1974, between Texaco International Petroleum Company and Government of Portugal	Supp. 34 p. 1
121	Contract of 19 December 1966 between Gulf Oil Corp. and the Government of Portugal for Development of Hydrocarbons in the Province of Cabinda	PL/S & CA/ BOL/CC Supp. XI A–0
122	Angol-Texaco Agreement for Off-shore Exploration, Decree 48/846 of 1 June 1967, including Appendix I, re Budgets; and Appendix II, re Accounting Principles of the Association	Supp. XXVI C–0
123	Text of Argo Petroleum Corporation Concession Contract, dated 13 April 1972: Appendix I – Partnership Contract between Concessionaire and State in accordance with Art. 15 of Concession Contract (provides for 30 per cent Government Participation); Appendix II – Technical Regulations regarding Collection of Samples	Supp. XXX B–0
124	Text of Sun Oil Group Concession Contract published in Official Gazette 21	Supp. 38 p. 1
125	Off-shore Concession granted 6 March 1974 to Esso Exploration Inc. (see subsequent Supplement for Text of Joint Venture Contract between Esso Exploration Inc. and Angola State Co.)	PL/S & CA/ BOL/CC Supp. 39 p. 1
126	Joint Venture Contract for Petroleum between Esso Exploration Production Angola Inc. and State Oil Company (forms part of Esso Off-shore Concession dated 6 March 1974)	Supp. 40 p. 1
127	Convention between Union Oil Company of California and the Government of the Republic of Dahomey, made and registered on 19 December 1964	Supp. XVIII A–0
128	Agreement between Ethiopian Government and Sinclair Oil Corp., dated 15 July 1945	Vol. I A–1
129	Agreement between Somali Territory and Sinclair Somali Corp., dated 8 May 1952	Vol. I A–1
130	Off-shore Oil Prospecting Licence (form offered by the Government of Ghana during 1968 for Off-shore Oil and Gas Rights)	Supp. XV A–0

271

S. No.	Agreement	Source
131	Concession Agreement between Esso Exploration Guine, Inc. and the Portuguese province of Guinea, 1958	Vol. I A–1
132	Convention signed 1973 between Government of Portugal and Esso Exploration Guine, Inc. Providing for 10 per cent State Participation (signed at same time as Esso Exploration Guine 1973, Off-shore Concession)	Supp. XXXV A–0
133	Standard Production-Sharing Contract, issued 1975	Supp. 41 p. 1
134	Memorandum of Agreement ("Government Agreement") between the Government of Lesotho on the one hand and Lesotho National Development Corp. (LNDC) and Westrans Industries Inc. on the other, dated 3 November 1972. Defines terms, including Income Tax, Royalty, Employment of Nationals, Exchange Controls, Import/Export Privileges, Marketing, Unitisation, Arbitration, Financial Records, etc. Includes First Schedule (determination of Taxable Income). Copies of Operating Agreement, Prospecting Lease, Mining Lease not included in this copy – see "Operating Agreement"	PL/S & CA/ BOL/CC Supp. XXXI A–0
135	Agreement between the Government of Liberia and J. J. Simmons, Jr, dated 18 February 1957	Supp. I A–0
136	Concession Agreement between the Government of the Republic of Liberia and Frontier Liberia Oil Company, Inc., dated 18 July 1969	Supp. XXVIII A–0
137	Convention of Establishment between the Islamic Republic of Mauritania and the Planet Oil & Minerals Corporation, dated 29 July 1966	Supp. XIX A–0
138	Mining Agreement between Islamic Republic of Mauritania and Texaco Mauritania Inc., dated 11 January 1971	Supp. XXVII B–0
139	Appendix to Statute No. 71081 and Establishment and Operating Agreement between Islamic Republic of Mauritania and Texaco Inc., dated 11 January 1971	Supp. XXVII C–0
140	Agreement between Mozambique Province, Mozambique Gulf Oil Co., and Mozambique Pan American Oil Co., 1958	Supp. I A–1
141	1968 Concession Contract between MOZGOC and PANAMOZ, subsidiaries of Gulf Oil Corp., and American International Oil Co.	Supp. XIII A–0
142	Tax Exemption for Oil Operations (see Europe – Basic Oil Laws and Concession Contracts – PORTUGAL – Supp. XXIV-A-0)	PL/EU/BOL/CC Supp. XXIV A–0
143	Etosha Petroleum Company Exploration Agreement of 13 October 1960 containing Prospecting and Mining Grant No. M.4/4/90	Supp. XXX A–0
144	Decree 63.154/MTP/M/U, Establishing a Model Convention for Petroleum Concessions	PL/S & CA/ BOL/CC Supp. IX A–0

272

S. No.	Agreement	Source
145	Decree 69-170/MTP/M/U of 5 December 1969, According Texaco Niger, Inc. an Exploration Permit for Hydrocarbons	Supp. XXVI A–0
146	Oil Exploration License between the Government of Nigeria on the one hand, and Shell Overseas Exploration Co. Ltd and D'Arcy Exploration Co. Ltd, on the other, dated 16 July 1949	Supp. VIII B–0
147	Oil Mining Lease between Shell-BP Petroleum Development Co. of Nigeria Ltd, and the Government of Nigeria	Supp. XIV A–0
148	Model Contract of May 1971 for Imposition of OPEC Conditions and Increased Posted Prices and Taxes signed by Nigerian Oil Operators in 1971	Supp. XXV A–0
149	Convention of 29 May 1961 between Somali Republic, Sinclair Somal Oil Corp., Amerada Petroleum, Continental Oil Co. of Somalia, and Ohio Oil International of Somalia	Supp. I B–1
150	Decree 62 of 15 March 1969 containing the Convention between the Government of the Somali Republic and the "Group of German Oil Companies in Somalia", relating to research and Exploration of Petroleum Resources	Supp. XXIII A–0
151	Agreement between the Republic of Sudan and AGIP Mineraria (Sudan) Ltd, for Exploration of Petroleum, dated 20 August 1959	Supp. VIII A–0
152	Petroleum Exploration Prospecting and Production Contracting Agreement between the Government of the United Republic of Tanzania and AGIP S.p.A., dated 14 April 1969	Supp. 38 p. 52
153	Agreement between Togoil Establishment and the Government of Togo for Exploration and Exploitation of Petroleum, dated 29 June 1966 (English and French texts)	Supp. XII A–0
154	Agreement between Union Oil Co. of California and the Government of Togo, for Exploration and Exploitation of Petroleum (English and French texts), signed by Union Oil Co. of California on 11 January 1966 (never ratified by the Government of Togo)	Supp. XII B–0
155	Model Petroleum Contract Agreement of 1972 issued under Art. 125 of the Constitution and Art. 3 of the Petroleum Code. Contains Form Contract for future Petroleum Exploration and Development Rights	PL/CA & C/. BOL/CC Supp. XV A–0
156	Concession Contract, dated 20 January 1972, between Wendell Phillips Oil Company and the Government of Haiti, covering On-shore and Off-shore Areas	Supp. XIII A–0
157	Agreement of 20 December 1966 between Signal Exploration (Jamaica) Company, and the Government of Jamaica for Exploration and Development of Off-shore Waters	Supp. IX A–0
158	Agreement between Petroleos Mexicanos and Edwin W. Pauley, Signal Oil and Gas Co., and American Independent Oil Co.	Supp. II A–1

274

S. No.	Agreement	Source
171	Contract for Petroleum Operations in the Jungle, between Petroleos del Peru, Peruvian Sun Oil Company Continental Oil Company of Peru, and Champlin Oil Peru, Inc., signed 2 September 1972	Supp. XXXIV C–0
172	Service Contracts – Minimum Bases as proposed by Venezuelan Petroleum Corporation (CVP), 13 March 1968, for Southern part of Lake Maracaibo	OPEC/Sel. Doc. IPI 1968 p. 381 *or* PL/SA/BOL/CC Supp. XX A–0
173	Maracaibo Service Contract between Occidental Petroleum de Venezuela and Corporacion Venezolana del Petroleo (CVP), dated 22 May 1971	Supp. XXXIII A–0

*Source:

The *Petroleum Legislation* series published from New York proved to be the most useful source for texts of agreements. Assistance received from Mr. Gordon Barrows, the Publisher, during the stage of collection of materials is gratefully acknowledged.

The abbreviations: PL/__/BOL/CC stand for: *Petroleum Legislation: Basic Oil Laws & Concession Contracts*. It has separate volumes, region-wise, thus: PL/*ME*/ BOL/CC refers to the Middle East (ME) region. The other regions are: NA – North Africa; A & A – Asia & Australasia; EU – Europe; S & CA – South and Central Africa; SA – South America; Supplements are issued region-wise.

Other sources consulted included:

OPEC/Sel. Doc. IPI: Selected Documents of the International Petroleum Industry, published by OPEC Secretariat, Vienna.

ILM: *International Legal Materials*, published by the American Society of International Law.

List of Legislation

Australia

Queensland
Petroleum Act, 1923 to 1958 (Consolidated Acts).
Act No. 26 of 1964.
The Petroleum (Submerged Lands) Act, 1967 (No. 36 of 1967).

New South Wales
Petroleum Act, 1955 (No. 28 of 1955).
Regulations to the Petroleum Act, 1955.
Petroleum (Submerged Lands) Taxation Act, 1967.

Western Australia
Petroleum Act, 1936 (No. 36 of 1936).
Regulations under the Petroleum Act, 1936.
Petroleum Amendment Act, 1940 (No. 8 of 1940).
Petroleum Amendment Act, 1949 (No. 25 of 1949).
Petroleum Amendment Act, 1951 (No. 12 of 1951).
Petroleum Amendment Act, 1954 (No. 66 of 1954).
Petroleum Amendment Act, 1966.
Petroleum (Submerged Lands) Act, 1967.
Petroleum Act, 1967.
Petroleum (Submerged Lands) Registration Fees Act, 1967 (No. 40 of 1967).
Petroleum Registration Fees Act, 1967 (No. 77 of 1967).

South Australia
Mining (Petroleum) Act, 1940–58.
Mining (Petroleum) Act, 1940.
Mining (Petroleum) Amendment Act, 1963.
Petroleum Act Amendment Act, 1969 (No. 90 of 1969).

Tasmania
Mining Act, 1929.
Marine Regulations, 1930.
Mining Act (Amending the Mining Act, 1929).
Mining Act, 1964.
Mining Act, 1966.
Petroleum (Submerged Lands) Act, 1967 (No. 63 of 1967).
Mining Act, 1972 (No. 17 of 1972).

Victoria
Petroleum Act, 1958.
Underseas Mineral Resources Act, 1963.
The Petroleum (Submerged Lands) Act, 1967.
Regulations Relating to Petroleum Exploration Permits, Petroleum Prospecting

276

Licences, and Petroleum Mineral Leases, published in Government Gazette 414 dated 20 April 1956.

Northern Territory
Petroleum (Prospecting and Mining) Ordinance, 1954–60.
Petroleum (Prospecting and Mining) Ordinance, 1964.
Regulations under the Petroleum (Prospecting and Mining) Ordinance, 1954–64, Regulations 1966, No. 6.
Regulations under the Petroleum (Prospecting and Mining) Ordinance, 1954–66, Regulations 1967, No. 10.

Bangladesh

The Bangladesh Petroleum Act, 1974.

Canada

Canada
National Energy Board Act, 1959.
Oil and Gas Production and Conservation Act, 1968–69.
Oil Export Tax Act, 1973–74.
Indian Oil and Gas Act, 1974.
Petroleum Administration Act, 1974–75.
Petro-Canada Act, 1975.
Petroleum and Natural Gas Act, 1976.

Alberta
Gas Utilities Act, 1970.
Mineral Taxation Act, 1970.
Oil and Gas Conservation Act, 1970.
Surface Rights Act, 1972.

British Columbia
Petroleum and Natural Gas Act, 1965.

Manitoba
Natural Resources Act, 1930.
Mines Act, 1970.

New Brunswick
Oil and Natural Gas Act, 1952.

Newfoundland
Petroleum and Natural Gas Act, 1965.

Nova Scotia
Petroleum and Natural Gas Act, 1967.

Ontario
Petroleum and Natural Gas Act, 1965.
Gas and Oil Leases Act, 1970.
Mining Act, 1970.

Prince Edward Island
Oil, Natural Gas and Minerals Act, 1957.

Quebec
Mining Act, 1965.

Saskatchewan
Natural Resources Act, 1930.

277

Mineral Contracts Re-negotiation Act, 1959.
Oil and Gas Conservation Act, 1965.
Surface Rights Acquisition Act, 1968.

India

Petroleum Act, 1934 (30 of 1934) (as modified up to 1 February 1958).
Oil and Natural Gas Concession Act, 1959.
Petroleum Concession Rules, 1949 (as amended through 1955).
Petroleum and Natural Gas Rules, 1959.
Petroleum and Natural Gas (Amendment) Rules, 1964.
Petroleum and Natural Gas (Amendment) Rules, 1965.
The Petroleum Pipelines (Acquisitions of Right of User in Land), Act, 1962 (No. 50 of 1962).

Malaysia

Continental Shelf Act, 1966 (No. 57 of 1966).
Petroleum Mining Act, 1966 (No. 58 of 1966).
Petroleum Mining Act, 1966 (No. 95 of 1972) superseding Petroleum Mining Act, 1966 (No. 58 of 1966).
Petroleum Mining Rules, 1968 (including Model Petroleum Agreement for Onshore and Offshire Operations).
Petroleum Income Tax Act (No. 45 of 1967).
Petroleum (Income Tax) (Amendment) Act (No. 79 of 1967).
    (Amends The Petroleum (Income Tax) Act (No. 47 of 1967).
Petroleum and Electricity (Control of Supplies) Act, 1974.
Petroleum Development Act, 1974.
Petroleum Development Amendment Act, 1975.

Sarawak
Oil Mining Ordinance, 1958.
The Oil Mining Ordinance, 1958.
The Sarawak (Definition of Boundaries) Order in Council, 1958.

North Borneo
Oil Mining Regulations, 1956.
The North Borneo (Alteration of Boundaries) Order in Council, 1954.
The North Borneo (Definition of Boundaries) Order in Council, 1958.
Petroleum Regulations, 1974.

New Zealand

Petroleum Act, 1937.
Petroleum Regulations, 1939.
Petroleum Amendment Act, 1953.
Petroleum Amendment Act, 1955.
Petroleum Amendment Act, 1962.
Continental Shelf Act, 1964.
Statutes Amendment Act, 1941.
New Zealand Territorial Sea and Fishing Zone Act, 1965.
The Petroleum Amendment Act, 1965.
The Petroleum Amendment Act 1967.
The Petroleum Amendment Act, 1974.
The Petroleum Amendment Act 1975.

Nigeria

Mineral Oils (Amendment) Ordinance, 1959.
Mineral Oils (Amendment) Ordinance, 1958.
Mineral Oils Ordinance, 1914.
Petroleum Ordinance, 1918.
Ordinance No. 55 of 1945 and Ordinance No. 35 of 1946.
Mineral Oils (Amendment) Ordinance, 1950.
Mineral (Amendment Regulations, 1950.
The Mineral and (Safety) Regulations, 1952
Petroleum Profits Tax Ordinance, 1959.
Mineral Oils Act No. 24 of 1962.
The Petroleum Profits Tax (Amendment) Decree, 1967.
Petroleum Profits Tax (Amendment) Decree, 1973.
Petroleum Decree No. 51 of 1969.
Petroleum (Drilling and Production) Regulations, 1969.
The Territorial Waters (Amendment) Decree, 1971.
The Nigerian National Oil Corporation Decree, 1971.
Offshore Oil Revenues Decree No. 9, 1971.
The Petroleum (Amendment) Act, 1974 (No. 38 of 1974).
Petroleum Refining Regulations, 1974.
Petroleum Equalization Fund (Management Board, etc.) Decree, 1975.

Norway

Royal Decree of 31 May 1963.
Act No. 12 of 21 June 1963, relating to Exploration for and Exploitation of Submarine Natural Resources.
Act No. 3 of 11 June 1965, relating to the Taxation of Submarine Petroleum Deposits
Royal Decree of 25 August 1967, relating to Safe Practice etc., in Exploration and Drilling for Submarine Petroleum Resources.
Royal Decree of 31 January 1969 relating to Scientific Research for Natural Resources on the Norwegian Continental Shelf, etc.
Royal Decree of 12 June 1970, relating to Provisional Rules concerning Exploration for certain Submarine Natural Resources other than Petroleum on the Norwegian Continental Shelf, etc.
Royal Decree of 8 December 1972, relating to Exploration for and Exploitation of Petroleum in the Seabed and Substrata of the Norwegian Continental Shelf.
Royal Decree of 14 February 1974, relating to the Taxation of Submarine Petroleum Deposits, etc.
Royal Decree of 3 October 1975, relating to Safe Practice, etc., in Exploration and Drilling for Submarine Petroleum Resources.
Royal Decree of 9 July 1976 relating to Safe Practice for the Production, etc., of Submarine Petroleum Resources.
Royal Decree of 9 July 1976 relating to Protection of Workers, etc., in Activities associated with Exploration and Exploitation of Submarine Petroleum Resources.

Tanzania

Ordinance 12 of 29 May 1958, to amend and consolidate the law relating to Mineral Oil.
The Mining (Mineral Oil) Regulations, 1958.

Trinidad

Land (Oil Mining) Regulations, 1934.

Submarine (Oil Mining) Regulations, 1945.
Land (Oil Mining) Regulations, 1949.
Submarine (Oil Mining) Amendment) Regulations, 1949.
Submarine (Oil Mining) Amendment Regulations, 1954.
The Continental Shelf Act, 1969.
The Petroleum Act, 1969 (No. 46 of 1969).
The Petroleum Regulations, 1970.

United Kingdom

Petroleum (Production) Act, 1918.
Petroleum (Production) Act, 1943.
Petroleum (Production) Regulations, 1935.
Gas Act of 1948.
Petroleum (Production) Amendment) Regulation, 1954.
Petroleum (Production) (Amendment) Regulations, 1957.
Pipelines Act, 1962.
Continental Shelf Act, 1964.
The Petroleum (Production) (Continental Shelf and Territorial Sea) Regulations, 1964.
The Continental Shelf (Designation of Area) Order, 1964.
The Continental Shelf (Jurisdiction) Order, 1964.
The Petroleum (Production) Regulations, 1966.
Pipe-Lines Act of 1962.
revention of Oil Pollution Act, 1971.
The Petroleum (Production) (Amendment) Regulations, 1971.
The Offshore Installations (Registration) Regulations, 1972.
The Offshore Installations (Managers) Regulations 1972.
Oil Taxation Act, 1975.
Petroleum and Submarine Pipelines Act, 1975.

Index